ANTARCTICA AND THE SECRET SPACE PROGRAM

David Hatcher Childress

Adventures Unlimited Press

Other Books by David Hatcher Childress:

VIMANA
ARK OF GOD
ANCIENT TECHNOLOGY IN PERU & BOLIVIA
THE MYSTERY OF THE OLMECS
PIRATES AND THE LOST TEMPLAR FLEET
TECHNOLOGY OF THE GODS
A HITCHHIKER'S GUIDE TO ARMAGEDDON
LOST CONTINENTS & THE HOLLOW EARTH
ATLANTIS & THE POWER SYSTEM OF THE GODS
THE FANTASTIC INVENTIONS OF NIKOLA TESLA
LOST CITIES OF NORTH & CENTRAL AMERICA
LOST CITIES OF CHINA, CENTRAL ASIA & INDIA
LOST CITIES & ANCIENT MYSTERIES OF AFRICA & ARABIA
LOST CITIES & ANCIENT MYSTERIES OF SOUTH AMERICA
LOST CITIES OF ANCIENT LEMURIA & THE PACIFIC
LOST CITIES OF ATLANTIS, ANCIENT EUROPE & THE MEDITERRANEAN
LOST CITIES & ANCIENT MYSTERIES OF THE SOUTHWEST
YETIS, SASQUATCH AND HAIRY GIANTS

With Brien Foerster
THE ENIGMA OF CRANIAL DEFORMATION

With Steven Mehler
THE CRYSTAL SKULLS

ANTARCTICA AND THE SECRET SPACE PROGRAM

Adventures Unlimited Press

Antarctica and the Secret Space Program

ISBN 978-1-948803-20-5

Published by:
Adventures Unlimited Press
One Adventure Place
Kempton, Illinois 60946 USA
auphq@frontiernet.net

AdventuresUnlimitedPress.com

10 9 8 7 6 5 4 3 2 1

One of the early German-Austrian designs for a working "flying saucer."

The logo of the Thule Society, founded in 1919.

ANTARCTICA AND THE SECRET SPACE PROGRAM

David Hatcher Childress

The specially-designed badge or emblem of the Deutsche Antarktische Expedition of 1938-1939. The Swastika and the oak leaves clearly reveal the involvement of the Thule Society.

TABLE OF CONTENTS

Rudolf Hess in a uniform that includes the oak leaves of the Thule Society.

Chapter One

The Mystery of Rudolf Hess

There are truly more things in Heaven
and in earth than man has dreamt.
—Karl Haushofer, 1943

Probably the most enigmatic of the postwar Nazis was Rudolf Hess: The second in line to succeed Hitler, after Herman Göring; a mystic and occultist who was part of the mysterious Thule Society; a pilot and mountain hiker; a man who was held incognito at Spandau Prison in Berlin by four different nations—the only prisoner in the facility from 1966 to 1987. What did Rudolf Hess know that kept him in prison for so many years at great financial cost? It was rumored that British Intelligence were responsible for Hess's supposed "suicide" in 1987. Why would they want to kill Hess after so many years? It has also been written that Hess was an early mind-control victim of the British and their mind-control doctors at the Tavistock Institute outside of London. This accounted for some of Hess's strange behavior during the Nuremberg Trials. It has also been suggested in several books that the man who died at Spandau Prison was not even the real Rudolf Hess. The more we look at Rudolf Hess the stranger and stranger this man and his story become.

Will the Real Rudolf Hess Please Stand Up?

Rudolf Hess was born in Alexandria, Egypt, on April 26, 1894, the son of a prosperous wholesaler and exporter. He came to Germany for the first time when he was fourteen. At the age of 20 he volunteered for the German Army at the outbreak of World War I in 1914. This was partly to escape the control of his domineering father who had refused to let him go to a university and instead wanted him to be part of the family business. Young Rudolf Hess had other ideas.

Hess was wounded twice during the war, and later became an airplane pilot. Hess was a large and powerful man, now a battle-hardened killer, and after the war he joined the Freikorps, a right-wing organization of ex-soldiers for hire. The Freikorps were involved in violently putting down Communist uprisings in Germany, often by having literal fistfights in the street.

Hess began attending the University of Munich where he studied political science. At the university he met Professor Karl Haushofer and joined the Thule Society, a secret society of sorts that espoused Nordic supremacy, mystical Germanic views of antiquity (such as a belief in Atlantis and Tibetan masters), and anti-Semitic views in the sense that the Jewish Old Testament was not the most important book in the world.

Professor Karl Haushofer was central to the Thule Society and was a former general whose theories on expansionism and race formed the basis of the concept of Lebensraum—increased living space for Germans at the expense of other nations. Haushofer's teachings were very influential in Germany's ultimate invasion of Eastern Europe under the belief that the Germans needed to expand their territory and culture. Their allies in Japan had a similar belief in

Karl Haushofer with Rudolf Hess.

Japan's alarming Imperial expansion throughout eastern Asia and the Western Pacific.

On July 1, 1920, Hess heard Adolf Hitler speak in a small Munich beer hall, and immediately joined the Nazi Party, becoming the sixteenth member. After his first meeting with Hitler, Hess said he felt "as though overcome by a vision." Hess was to become utterly devoted to Hitler and eagerly agreed with everything the shouting politician said. At early Nazi Party rallies, Hess was a formidable fighter who continually brawled with Marxist activists and others who violently attempted to disrupt Hitler's speeches.

In 1923, Hess took part in Hitler's failed Beer Hall Putsch. Hitler and the Nazis attempted to seize control of the government of Bavaria at this time and Hess was arrested and imprisoned along with Hitler at Landsberg Prison. While the two were in prison, Hess took dictation for Hitler's book, *Mein Kampf*. Hess also made some editorial suggestions regarding the organization of the Nazi Party, the notion of Lebensraum, plus material in the book about the historical role of the British Empire.

Both Hitler and Hess were released from prison in 1925 and Hess served for several years as the personal secretary to Hitler in spite of having no official rank in the Nazi Party. In 1932, Hitler appointed Hess an SS General and the Chairman of the Central Political Commission of the Nazi Party as a reward for his loyal service. On April 21, 1933, Hess was made Deputy Führer, a figurehead position with mostly ceremonial duties.

It is said that Hess was a shy, insecure man who displayed "fanatical loyalty and absolute blind obedience" to Hitler. Hess gave a revealing speech in 1934, stating:

> With pride we see that one man remains beyond all criticism, that is the Führer. This is because everyone feels and knows: he is always right, and he will always be right. The National Socialism of all of us is anchored in uncritical loyalty, in the surrender to the Führer that does not ask for the why in individual cases, in the silent execution of his orders. We believe that the Führer is obeying a higher call to fashion German history. There can be no criticism of this belief.

Hess would become famous for being the speaker to announce the Führer at mass meetings with bellowing, wide-eyed fanaticism, as can be seen in the Nazi documentary *Triumph of the Will*.

The Nazi regime began to persecute Jews soon after its seizure of power. Hess's office was partly responsible for drafting Hitler's Nuremberg Laws of 1935, laws that had far-reaching implications for the Jews of Germany, banning marriage between non-Jewish and Jewish Germans and depriving non-Aryans of their German citizenship. Hess's friend Karl Haushofer and his family were affected by these laws, as Haushofer had married a half-Jewish woman, so Hess issued documents exempting them from this legislation.

Hess was motivated by his loyalty to Hitler and a desire to be useful to him; he did not seek power or prestige or take advantage of his position to accumulate personal wealth. He and his wife Ilse lived in a modest house in Munich. Although Hess had less influence than other top NSDAP officials, he was popular with the masses.

Hess and Neuschwabenland

In December of 1938 Hermann Göring launched an expedition to Antarctica in an effort to establish a naval base and whaling station on the polar continent. The New Swabia Expedition left Hamburg for Antarctica aboard *MS Schwabenland* on December 17, 1938. The *MS Schwabenland* was a freighter built in 1925 and renamed in 1934 after the Swabia region in southern Germany. The *MS Schwabenland* was also able to carry special aircraft that could be catapulted from the deck.

The expedition was top secret and was overseen by Göring himself. The Thule Society was also apparently involved. The expedition had 33 members plus the *Schwabenland*'s crew of 24. On January 19, 1939 the ship arrived at the Princess Martha Coast of Antarctica, in an area which had recently been claimed by Norway as Queen Maud

The emblem of the expedition.

The German ship *Schwabenland*.

Land, and began charting the region. Nazi German flags were placed on the sea ice along the coast. Naming the area Neuschwabenland after the ship, the expedition established a temporary base, and in the following weeks teams walked along the coast recording claim reservations on hills and other significant landmarks.

Some researchers, such as Joseph Farrell, believe that Hess was part of this expedition to Antarctica. Upon his return in April of 1939 Hess reported to Göring and Hitler what had been discovered in Antarctica and whether they had found a suitable place for a German naval base. Such a base would serve commercial ships such as whalers as well as military ships and submarines.

This success must have been one of the reasons that Hitler made Hess second in line to succeed him, after Hermann Göring, in September of 1939. During this same time, Hitler appointed Hess's chief of staff, Martin Bormann, as his personal secretary, a post formerly held by Hess. The German invasion of Poland happened at this time as well.

That Antarctica was part of Göring, Hess and Haushofer's plan for German expansion and the recreation of a global German colonial community is quite clear. Joseph Farrell quotes the German author Heinz Schön, who wrote a 2004 book in German called *Mythos Neuschwabenland: Für Hitler am Südpol: Die deutsche Antarktis-expedition 1938/39* and says:

> As commissioner for the Four-Year Plan, Göring knew the importance for Germany of whaling in Antarctica, and how essential it was to ensure this, and to open up new fishing grounds. It seemed high time for him to send a large expedition to Antarctica. On May 9, 1938, a plan for an Antarctic expedition, drawn up by the staff of his ministry, which was to be carried out in the Antarctic summer of 1938/39, was presented to him. He approved, and commissioned Helmut Wohlthat as Minister-Director for special projects, with the preparation of the expedition, and conferred upon him all his powers of authority.[2]

The New Swabia Expedition was the third German expedition to Antarctica. The first German expedition to Antarctica was the Gauss expedition from 1901 to 1903. Led by Arctic veteran and geologist Erich von Drygalski, this was the first expedition to use a hot air balloon in Antarctica. The expedition also found and named Kaiser Wilhelm II Land. The second German Antarctic expedition was from 1911 to 1912. Led by Wilhelm Filchner it had a goal of crossing Antarctica to learn if it was one piece of land. This crossing of the icy continent never materialized, but the expedition discovered and named the Luitpold Coast and the Filchner Ice Shelf.

Decades went by and then a German whaling fleet was put to sea in 1937 and, upon its successful return in early 1938, plans for the third German Antarctic expedition were drawn up. The third German Antarctic Expedition (1938–1939) was led by Alfred Ritscher (1879–1963), a captain in the German Navy. As noted above, one of main purposes of the secret expedition was to find an area in Antarctica for a German whaling station, as a way to

A seaplane launches off of the German ship *Schwabenland*.

A Nazi flag ceremony in Neuschwabenland.

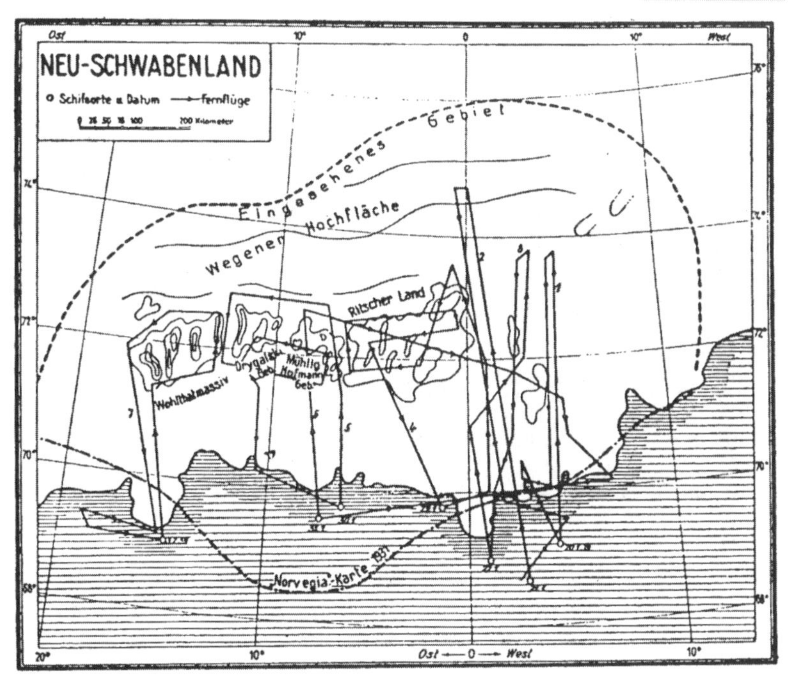

A map of the flag planting and flights over Neuschwabenland.

Two ice caves explored in the 1938-39 Antarctic Expedition.

increase Germany's production of fat. Whale oil was then the most important raw material for the production of margarine and soap in Germany and the country was the second largest purchaser of Norwegian whale oil, importing some 200,000 metric tons annually.

Germany did not want to be dependent on imports and it was

thought that Germany would soon be at war, which would put a lot of strain on Germany's foreign currency reserves. The other goal of this secret expedition was to scout possible locations for a German naval base and that would include a base for submarines.

While in Antarctica, seven photographic survey flights were made by the ship's two Dornier Wal seaplanes named *Passat* and *Boreas*. About a dozen 1.2-meter (3.9 foot)-long aluminum arrows, with 30-centimeter (12 inch) steel cones and three upper stabilizer wings embossed with swastikas, were airdropped onto the ice at turning points of the flight polygons (these arrows had been tested on Austria's Pasterze glacier before the expedition). Supposedly, none of these have ever been recovered.

Eight more flights were made to areas of keen interest and on these trips, some of the photos were taken with color film. Altogether they flew over hundreds of thousands of square kilometers and took more than 16,000 aerial photographs, some of which were published after the war by Alfred Ritscher.

During one flight, the ice-free Schirmacher Oasis was spotted from the air by Richard Schirmacher (who named it after himself) shortly before the *Schwabenland* left the Antarctic coast on February 6, 1939. On its return trip to Germany the expedition made oceanographic studies near Bouvet Island and Fernando de Noronha, and arrived back in Hamburg on April 11, 1939. Meanwhile, the

An ice cave explored in the 1938-39 Antarctic Expedition.

Norwegian government had learned about the secret expedition through reports from whalers who had been along the coast of Queen Maud Land. Germany issued a decree about the establishment of a German Antarctic Sector called New Swabia after the expedition's return in April of 1939.

Hess and the Flight to Scotland

It is said that Hess was obsessed with his health to the point of hypochondria, consulting many doctors and other practitioners for what he described to his British captors as a long list of ailments involving the kidneys, colon, gall bladder, bowels and his heart. Hess was a vegetarian, like Hitler and Himmler, and he did not smoke or drink. He was a big believer in homeopathic medicines and Rudolf Steiner-type food that was "biologically dynamic." Hess was interested in music, enjoyed reading and loved to spend time hiking and climbing in the mountains with his wife, Ilse. He and his friend Albrecht Haushofer, Karl Haushofer's son, shared an interest in astrology, psychic powers, clairvoyance and the occult.

As the war progressed, Hitler's attention became focused on foreign affairs and the conduct of the war. Hess, who was not directly engaged in these endeavors, became increasingly sidelined from the affairs of the nation and from Hitler's attention. Martin Bormann had successfully supplanted Hess in many of his duties and essentially usurped the position at Hitler's side that Hess had once held. Hess later said that he was concerned that Germany would face a war on two fronts as plans progressed for Operation Barbarossa, the invasion of the Soviet Union scheduled to take place in 1941.

Hess decided to attempt to bring Britain to the negotiating table by travelling there himself to seek direct meetings with the members of the British government. He asked the advice of Albrecht Haushofer who suggested several potential contacts in Britain. Hess settled on fellow aviator Douglas Douglas-Hamilton, the Duke of Hamilton, whom he had met briefly during the Berlin Olympics in 1936.

On Hess's instructions, Haushofer wrote to Hamilton in September of 1940, but the letter was intercepted by MI5 and Hamilton did not see it until March of 1941. The Duke of Hamilton was chosen because he was one of the leaders of an opposition party that was opposed to war with Germany, and because he was a friend of Haushofer. In a letter that Hess wrote to his wife dated November 4, 1940, he says that in spite of not receiving a reply from Hamilton, he intended to proceed with his plan to fly himself to Scotland and meet with Hamilton.

Hess began training on the Messerschmitt 110, a two-seater twin-engine aircraft, in October of 1940 under the chief test pilot at Messerschmitt. He continued to practice, including logging many cross-country flights, and found a specific aircraft that handled well—a Bf 110E-1/N—which was from then on held in reserve for his personal use. He asked for a radio compass, modifications to the oxygen delivery system, and large long-range fuel tanks to be installed on this plane, and these requests were granted in March 1941.

Hoping to save the Reich from disaster and redeem himself in the eyes of his Führer, Hess put on a Luftwaffe uniform and leather jacket (he was an SS General as well) and flew the fighter plane alone toward Scotland on a 'peace' mission, on May 10, 1941, just before the Nazi invasion of the Soviet Union. After a final check of the weather reports for Germany and the North Sea, Hess took off at 17:45 from the airfield at Augsburg-Haunstetten in his specially prepared aircraft.

Hess, Heinrich Himmler, Phillip Bouhler, Fritz Todt, Reinhard Heydrich, and others listening to Konrad Meyer at a Generalplan Ost exhibition, 1941.

It was the last of several attempts by Hess to fly to Scotland, with previous efforts having to be called off due to mechanical problems or poor weather. Wearing the leather flying suit, he brought along a supply of money and toiletries, a flashlight, a camera, maps and charts, and a collection of 28 different medicines, as well as dextrose tablets to help ward off fatigue, and an assortment of homeopathic remedies.

Hess flew north and when he reached the west coast of Germany near the Frisian Islands, he turned and flew in an easterly direction for twenty minutes to stay out of range of British radar. He then took a heading of 335 degrees for the trip across the North Sea, initially at low altitude, but travelling for most of the journey at 5,000 feet (1,500 meters). At 20:58 he changed his heading to 245 degrees, intending to approach the coast of northeast England near the town of Bamburgh, Northumberland.

As Hess approached the coast he realized that sunset was still nearly an hour away and he needed darkness to fly past the coast, Hess backtracked, zigzagging back and forth for 40 minutes until it grew dark. Around this time his auxiliary fuel tanks were exhausted, so he released them into the sea. Shortly after that he was over Scotland and at 6,000 feet Hess bailed out and parachuted safely to the ground where he encountered a Scottish farmer and told him in English, "I have an important message for the Duke of Hamilton."

Now in captivity, Hess told his captors that he wanted to convince the British government that Hitler only wanted Lebensraum for the German people and had no wish to destroy a fellow 'Nordic' nation. He also knew of Hitler's plans to attack the Soviet Union and wanted to prevent Germany from getting involved in a two-front war, fighting the Soviets to the east and Britain and its allies in the west.

During interrogation in a British Army barracks, Hess proposed that if the British would allow Nazi Germany to dominate Europe, then the British Empire would not be further molested by Hitler. Hess demanded a free hand for Germany in Europe and the return of former German colonies as compensation for Germany's promise to respect the integrity of the British Empire. Hess insisted that German victory was inevitable and said that the British people would be starved to death by a Nazi blockade around the British

Isles unless they accepted his generous peace offer.

Before his departure from Germany, Hess had given his adjutant, Karlheinz Pintsch, a letter addressed to Hitler that detailed his intentions to open peace negotiations with the British. Pintsch delivered the letter to Hitler at the Berghof (Hitler's home in the Bavarian Alps where he spent a great deal of time during WWII and became an important center of government) around noon on May 11. After reading the letter, Hitler let loose an angry yell that was heard throughout the entire Berghof, and sent for a number of his inner circle as he was concerned that a putsch (an attempt to overthrow the government) might be underway.

Hess's odd flight out of Germany, but not his destination or fate, was first announced by Munich Radio in Germany on the evening of May 12. On May 13 Hitler sent Foreign Minister Joachim von Ribbentrop to give the news in person to Mussolini, and on the same day the British press was permitted to release full information about the events. Ilse Hess finally learned that her husband had survived the trip when news of his fate was broadcast on German radio on May 14. Hitler publicly accused Hess of suffering from "pacifist delusions."

Hitler was worried that his allies, Italy and Japan, would perceive Hess's act as an attempt by Hitler to secretly open peace negotiations with the British. Hitler contacted Mussolini specifically to reassure him otherwise. Hitler ordered that the German press should characterize Hess as a madman who made the decision to fly to Scotland entirely on his own, without Hitler's knowledge. Subsequent German newspaper reports described Hess as "deluded, deranged," indicating that his mental health had been affected by injuries sustained during World War I.

Some members of the government, such as Göring and Propaganda Minister Joseph Goebbels, believed this only made matters worse, because if Hess truly were mentally ill, he should not have been holding such an important government position, second in line to succeed the Führer.

Hitler stripped Hess of all of his party and state offices, and secretly ordered him shot on sight if he ever returned to Germany. He abolished the post of Deputy Führer, assigning Hess's former duties to Bormann, with the title of Head of the Party Chancellery.

Bormann used the opportunity afforded by Hess's departure to secure significant power for himself. Meanwhile, Hitler initiated Aktion Hess, a flurry of hundreds of arrests of astrologers, faith healers and occultists that took place around June 9 and 10. The campaign was part of a propaganda effort by Goebbels and others to denigrate Hess and to make scapegoats of occult practitioners.

This process involved rounding up and imprisoning Hess's associates, including his wide-ranging network of occultists, astrologers, and ritualists. By positioning himself squarely at the center of the occult movement and then falling from grace so spectacularly, Hess doomed his fellow practitioners to a very sudden end. Everything from fortune-telling to astrology was outlawed, and the Nazi party's infatuation with the occult was over. However, the occult undercurrent remained in the SS, of which Hess had been a General.

Hitler has often been seen as the subject of various prophecies and the Nazis had sought to use the prophecies of Nostradamus for their own benefit. Perhaps most notable is the Nostradamus verse in which he wrote, "Beasts ferocious with hunger will cross the rivers; the greater part of the battlefield will be against Hister," along with references to a "Child of Germany."

One man who believed in Hitler's almost mythical status was a college professor named Johann Dietrich Eckhart. He was a member of the mysterious Thule Society, and he and many of the group's members believed that a German messiah was prophesied to enter history in the near future. This German leader would return the nation to its former glory, and avenge its defeat in the First World War, undoing the humiliation imposed upon the country with the Treaty of Versailles. Eckhart was a student of eastern mysticism and developed an ideology of a "genius superman," based on writings by the Völkisch author Jörg Lanz von Liebenfels. Eckhart met Hitler in 1919 and was certain that this man was the savior he believed Germany had been promised. The man went on to shape Hitler's ideologies considerably, sculpting the beliefs and worldview of the Nazi party. Eckhart died in 1923.

One person arrested during Aktion Hess was Karl Ernst Krafft, an astrologer and psychic who claimed he was clairvoyant. He was a committed supporter of the Nazi regime, but in 1939 he made a prediction of an assassination attempt against Adolf Hitler between the 7th and 10th of November of that year.

At the time his claims received little attention, but following the detonation of a bomb in the Munich Beer Hall on November 8, everything changed. Hitler had already left the building by the time the explosion occurred. It killed seven people and injured almost 70 more, but the target of the attack escaped unscathed. Soon afterward, word of Krafft's prophecy reached Rudolf Hess, and the fortune-teller was arrested. However, he managed to convince his interrogators that he was innocent of any wrongdoing and that his gifts of prophecy were genuine.

Krafft was well-liked by Hitler himself, and was ordered to begin an evaluation of the prophecies of Nostradamus that would favor the Nazi worldview. However, his own gifts were his undoing; as mentioned above, following Rudolf Hess's flight to Scotland, he was swept up in Aktion Hess. Krafft was arrested in 1941 and died in prison in 1945.

Back in England, Hess was being held as a prisoner of war.

American journalist Hubert Renfro Knickerbocker, who had met both Hitler and Hess, speculated that Hitler had sent Hess to deliver a message informing Winston Churchill of the forthcoming invasion of the Soviet Union, and offering a negotiated peace or even an anti-Bolshevik partnership. Soviet leader Joseph Stalin believed that Hess's flight had been engineered by the British. Stalin persisted in this belief as late as 1944, when he mentioned the matter to Churchill, who insisted that they had no advance knowledge of the flight. While some sources reported that Hess had been on an official mission, Churchill later stated in his book *The Grand Alliance* that in his view, the mission had not been authorized. "He came to us of his own free will, and, though without authority, had something of the quality of an envoy," said Churchill, and referred to Hess's plan as one of "lunatic benevolence."

After the war, Albert Speer discussed the rationale for the flight with Hess, who told him that "the idea had been inspired in him in a dream by supernatural forces. We will guarantee England her empire; in return she will give us a free hand in Europe."

When captured in Scotland, Hess was taken to Buchanan Castle where they discovered that they had captured a high-ranking Nazi.

From Buchanan Castle, Hess was transferred briefly to the Tower of London and then to Mytchett Place in Surrey, a fortified mansion, designated "Camp Z," where he was debriefed over the next year. Churchill issued orders that Hess was to be treated well, though he was not allowed to read newspapers or listen to the radio. By early June, Hess was allowed to write to his family. He also prepared a letter to the Duke of Hamilton, but it was never delivered, and his repeated requests for further meetings were turned down.

Psychiatrists who treated Hess during this period noted that while he was not insane, he was mentally unstable, with tendencies toward hypochondria and paranoia. Hess continually claimed he was being poisoned and was being prevented from sleeping. He would insist on swapping his dinner with that of one of his guards, and attempted to get them to send samples of the food out for analysis.

Spandau Prison and the Mystery of Prisoner Number 7

During his years of British imprisonment, Hess is said to have displayed increasingly unstable behavior and developed a paranoid obsession that his food was being poisoned. In 1945, Hess was returned to Germany to stand trial before the International Military Tribunal at Nuremberg.

Beginning in November of 1945 the Allies of World War II held a series of military tribunals and trials in Nuremberg, beginning with a trial of the major war criminals. Hess was tried with this first group of 23 defendants, all of whom were charged with four counts—conspiracy to commit crimes, crimes against peace, war crimes and crimes against humanity, in violation of international laws governing warfare.

One of the other defendants was Joachim von Ribbentrop, who was the Foreign Minister of Nazi Germany from 1938 until 1945. Ribbentrop first came to Adolf Hitler's notice as a well-travelled businessman with more knowledge of the outside world than most senior Nazis and as an authority on world affairs. Ribbentrop offered his house Schloss Fuschl for the secret meetings in January 1933 that resulted in Hitler's appointment as Chancellor of Germany. Before World War II, he played a key role in brokering the Pact of Steel (an alliance with Fascist Italy) and the Molotov–Ribbentrop Pact (the Nazi–Soviet non-aggression pact). Ribbentrop favored retaining good relations with the Soviets, and was opposed to the invasion of the Soviet Union.

Arrested in June of 1945, Ribbentrop was tried at the Nuremberg trials and convicted for his role in starting World War II in Europe and enabling the Holocaust. On October 16, 1946, he became the first of those sentenced to death by hanging and was soon executed.

On his arrival in Nuremberg, Hess was reluctant to give up some of his possessions, including samples of food he said had been poisoned by the British; he proposed to use these for his defense during the trial. In the courtroom, he suffered from spells of disorientation, staring off vacantly into space, and for a time claimed to have amnesia.

The chief psychiatrist at Nuremberg, Douglas Kelley of the US Military, gave the opinion that the defendant suffered from "a true psychoneurosis, primarily of the hysterical type, engrafted on a basic paranoid and schizoid personality, with amnesia, partly genuine and partly feigned," but found him fit to stand trial. Efforts were made to trigger his memory, including bringing in his former secretaries and showing old newsreels, but he persisted in showing no response to these stimuli.

In periods of lucidity Hess continued to display loyalty to Hitler, ending with his final speech:

> It was granted me for many years to live and work under the greatest son whom my nation has brought forth in the thousand years of its history. Even if I could I would not expunge this period from my existence. I regret nothing. If I were standing once more at the beginning I should act once again as I did then, even if I knew that at the end I should be burnt at the stake…

In spite of his mental condition, he was sentenced to life in Spandau prison in Berlin. The Soviets had wanted Hess to get the death sentence and blocked all attempts at early release. From 1966 to his death in 1987 Hess was the only inmate at the prison. He was designated Prisoner Number 7.

His fellow inmates—Konstantin von Neurath, Walther Funk, and Erich Raeder—were released because of poor health in the early 1950s; Admiral Karl Dönitz was released in 1956, and then Baldur von Schirach and Albert Speer were released in 1966.

Visitors were allowed to come for half an hour per month, but Hess forbade his family to visit until December 1969, when he was a patient at the British Military Hospital in West Berlin for a perforated ulcer. By this time his son, Wolf Rüdiger Hess, was 32 years old and his wife, Ilse, was 69 and they had not seen Hess since his departure from Germany in 1941. After this illness, he allowed his family to visit regularly.

It is said that Hess would cry out in the night, claiming he had stomach pains. He continued to suspect that his food was being poisoned and complained of amnesia. A psychiatrist who examined him in 1957 deemed he was not ill enough to be transferred to a mental hospital. Hess made a suicide attempt in 1977.

Other than his stays in hospital, Hess spent the rest of his life in

Rudolf Hess in Spandau Prison.

Spandau Prison. The 600-cell prison continued to be maintained for its lone prisoner from 1966 until Hess's death in 1987. Hess was eventually allowed to move fairly freely around the cell block, setting his own routine and choosing his own activities, which included television, films, reading, and gardening.

Hess's lawyer Alfred Seidl launched numerous appeals for his release, beginning as early as 1947. These were denied, mainly because the Soviets repeatedly vetoed the proposal. Spandau was located in West Berlin, and its existence gave the Soviets a foothold in that sector of the city. Additionally, Soviet officials believed, probably rightly, that Hess must have known in 1941 that an attack on their country was imminent.

In 1967, Wolf Rüdiger Hess began a campaign to win his father's release, garnering support from notable politicians such as Geoffrey Lawrence, 1st Baron Oaksey in Britain and Willy Brandt in Germany, but to no avail, in spite of the prisoner's advanced age and deteriorating health.

Hess died at age 93 on August 17, 1987, in a summerhouse that had been set up in the prison garden as a reading room. He had apparently taken an extension cord from one of the lamps, strung it over a window latch, and hanged himself. A short note to his family was found in his pocket, thanking them for all that they had done.

The Four Powers released a statement on September 17, ruling the death a suicide. Hess was buried in Wunsiedel, Bavaria, and his grave later became a pilgrimage site for neo-Nazis. In 2011 it was decided that his body should be moved. Hess's remains were subsequently cremated, and his ashes were scattered in an unidentified lake.

Immediately there was suspicion that Hess had been murdered at the prison. According to an investigation by the British government in 1989, the available evidence did not back up the claim that Hess was murdered, and Solicitor General Sir Nicholas Lyell saw no grounds for further investigation.

Hess's lawyer, Alfred Seidl, felt that Hess was too old and frail to have managed to kill himself. Wolf Rüdiger Hess repeatedly claimed that his father had been murdered by the British Secret Intelligence Service to prevent him from revealing information about British misconduct during the war.

According to Wikipedia, a report released in 2012 led to questions again being asked as to whether Hess had been murdered. In that year historian Peter Padfield claimed that the suicide note found on the body appeared to have been written when Hess was hospitalized in 1969, nearly 20 years before his final "suicide."

Why would Hess have been murdered? If the British had murdered Hess, why had they waited so long to kill him? Had Hess been an early mind-control subject of the British, causing part of his memory to be erased and causing his strange spells of amnesia? Joseph Farrell says that Allen Foster Dulles of the American OSS suspected as much when he watched Hess at the Nuremberg trials.[2]

There also arose speculation that the prisoner at Spandau prison was not actually Rudolf Hess at all but a replacement—someone also under mind control. This would be one possible reason for murdering Hess in prison—to keep the secret from getting out that this was not the real Rudolf Hess.

This myth that Spandau prisoner number 7 was not actually Hess was disproved in 2019, when a study of DNA testing undertaken by Sherman McCall, formerly of the Walter Reed Army Medical Center, and Jan Cemper-Kiesslich of the University of Salzburg demonstrated a 99.99 per cent match between the prisoner's y chromosome DNA markers and those of a living male Hess relative.

So, it appears that the prisoner at Spandau was the real Hess after all. But the whole undertaking shows that there was enough interest—on an international scale—to find out if Hess had been murdered and whether the prisoner kept alone in Spandau all of those lonely years was indeed the former SS General himself.

Why the intense interest in Hess? Was it only the Soviets that wanted to keep Hess locked up for decades? Had the British somehow altered Hess's personality during his captivity or was he just an unbalanced occultist who had a number of odd beliefs? What did Hess know about Antarctica and what had he told the British?

As we are about to see, the end of World War II brought to light a number of oddities that are little understood even today. These include missing submarines, secret bases in Tibet and Antarctica, flying saucers, suppressed technology and much more!

Chapter Two

The Mystery of James Forrestal

Go to Heaven for the climate,
Hell for the company.
—Mark Twain

We turn now to another curious figure in WWII, James Vincent Forrestal (February 15, 1892—May 22, 1949). Forrestal was the last Cabinet-level United States Secretary of the Navy and the first United States Secretary of Defense. He was nominated to be Undersecretary of the Navy by President Franklin D. Roosevelt in 1940, and he led the national effort for industrial mobilization for the war effort during World War II. He was named Secretary of the Navy in May 1944, and was the first Secretary of the newly created Defense Department in 1947 by Roosevelt's successor Harry S. Truman. Forrestal signed the order for Operation Highjump and the invasion of Antarctica. Forrestal was also supposedly a member of the Top Secret government control group known as MJ-12. He died under mysterious circumstances in 1949.

Forrestal was a big supporter of naval battle groups that were centered around aircraft carriers. As Secretary of Defense, it is said that he was often at odds with President Truman over national policy. As the Defense Department drew down after the war Forrestal fought for all the money he could get to keep the American Army from shrinking.

In 1948 Tomas Dewey was widely expected to win the Presidential election, and it was later revealed that Forrestal had met and negotiated for a cabinet position with Dewey, who was Truman's opponent. But Truman won the election and Forrestal was forced to resign as Secretary of Defense by the president in 1949. Shortly after his resignation he underwent medical care for depression, and died after falling from a sixteenth floor window of the hospital where he was being treated.

Forrestal was born in Matteawan, New York, the son of Irish immigrants. His mother raised him as a devout Roman Catholic. He was an amateur boxer. After graduating from high school at the age of 16, in 1908, he spent the next three years working for a trio of newspapers: the *Matteawan Evening Journal,* the *Mount Vernon Argus* and the *Poughkeepsie News Press*.

Forrestal entered Dartmouth College in 1911, but transferred to Princeton University in his sophomore year. He served as an editor for *The Daily Princetonian.* In 1926 Forrestal married Josephine Stovall a *Vogue* writer.

Forrestal went to work as a bond salesman in 1916. When the United States entered World War I, he enlisted in the Navy and eventually became a Naval Aviator, training with the Royal Flying Corps in Canada. During the final year of the war, Forrestal spent much of his time in ,Washington, DC, at the office of Naval Operations while completing his flight training, and eventually reached the rank of Lieutenant.

After the war, Forrestal returned to working in finance and made his fortune on Wall Street. He also acted as a publicist for the Democratic Party committee in Dutchess County, New York helping politicians from the area win elections at both the state and national level. One of those individuals aided by Forrestal's political work was a neighbor and fellow Democrat, Franklin D. Roosevelt.

When Franklin D. Roosevelt became president he appointed Forrestal as a special administrative assistant on June 22, 1940.

James Forrestal.

Six weeks later, he nominated him for the newly established position, Undersecretary of the Navy. In his nearly four years as undersecretary, Forrestal proved highly effective at mobilizing domestic industrial production for the war effort.

In September of 1942, in order to get a grasp on the reports for materiel being asked for by the Pacific fleet, he made a tour of naval operations in the Southwest Pacific and made a stop at Pearl Harbor. He became Secretary of the Navy on May 19, 1944, after his immediate superior Secretary Frank Knox died from a heart attack. Forrestal led the Navy through the closing year of the war and the painful early years of demobilization that followed. As Secretary, Forrestal introduced a policy of racial integration in the Navy.

Forrestal traveled to combat zones to see naval forces in action. He was present at the Battle of Kwajalein in 1944, and (as Secretary) witnessed the Battle of Iwo Jima in 1945. After five days of pitched battle, a detachment of Marines was sent to hoist the American flag on the 545-foot summit of Mount Suribachi on Iwo Jima. This was the first time in the war that the US flag had flown on Japanese soil. Forrestal, who had just landed on the beach, claimed the historic flag as a souvenir. A second, larger flag was run up in its place, and this second flag-raising was the moment captured by Associated Press photographer Joe Rosenthal in his famous photograph.

Forrestal, along with Secretary of War Henry Stimson and Undersecretary of State Joseph Grew, in the early months of 1945, strongly advocated a softer policy toward Japan that would permit a negotiated armistice, a 'face-saving' surrender. Forrestal's primary concern was not the resurgence of a militarized Japan, but rather "the menace of Russian Communism and its attraction for decimated, destabilized societies in Europe and Asia." Forrestal was for keeping the Soviet Union out of the war with Japan so the nation could not overly exert its influence in the area.

Forrestal's counsel on Japan was finally followed, but not until atomic bombs had been dropped on Hiroshima and Nagasaki. The day after the Nagasaki attack, the Japanese sent out a radio transmission saying that they were ready to accept the terms of the Allies' Potsdam Declaration, "with the understanding that said declaration does not comprise any demand which prejudices the prerogatives of His Majesty as a sovereign ruler."

The US had insisted on "unconditional surrender" up to this time, retaining the sticking point that had held up the war's conclusion for months. Strong voices within the administration, including Secretary of State James Byrnes, counseled fighting on. Forrestal came up with a shrewd and simple solution: Accept the offer and declare that it accomplishes what the Potsdam Declaration demanded. Say that the Emperor and the Japanese government will rule subject to the orders of the Supreme Commander for the Allied Powers. This would imply recognition of the Emperor while tending to neutralize American public passions against the Emperor. Truman liked this. It would be close enough to "unconditional."

Strongly anti-Soviet and anti-Communist, after the war, Forrestal urged President Truman to take a hard line with the Soviets over Poland. He also strongly influenced the new Wisconsin Senator, Joseph McCarthy. Upon McCarthy's arrival in Washington in December of 1946, Forrestal invited him to lunch and shared with him his concerns about Communist infiltration of

James Forrestal and Harry Truman.

the US, including the government.

In early 1946 Forrestal authorized the invasion of Antarctica, and Operation Highjump came into effect. To be discussed in the next chapter, Operation Highjump was a United States Navy operation organized by Rear Admiral Richard E. Byrd, Jr., and led by Rear Admiral Richard H. Cruzen. Operation Highjump commenced on August 26, 1946 and ended in late February of 1947. "Task Force 68" included 4,700 men, 13 ships, and 33 aircraft. Operation Highjump's primary mission was said to be the establishment of the Antarctic research base Little America IV, but stories persist that Operation Highjump was a military invasion of Antarctica that has never been fully explained.

1947: The Creation of Majestic-12

The next major thing that seems to have happened to Forrestal is that he became a founding member of a group of twelve men who were operating a committee to investigate alien spacecraft per an executive order from President Truman issued on September 24, 1947. This secret committee of scientists, military leaders, and government officials was to be called by a code name: Majestic 12 (or MJ-12).

During the early 1980s, several books were published concerning a cover-up of the Roswell UFO incident of July 1947; the authors speculated some secretive upper tier of the United States government or military was responsible. Then, in 1984 UFO researcher Jaime Shandera received an envelope containing film that, when developed, showed images of eight pages of documents that appeared to be briefing papers describing "Operation Majestic-12." The documents purported to reveal a secret committee of 12 scientists and military officers that existed in 1947. This committee was to look into how the recovered alien technology could be exploited, and how the US should engage with extraterrestrial life in the future.

Shandera, with his colleagues Bill Moore and Stanton T. Friedman, said they later received a series of anonymous messages that led them to find another document that has been called the "Cutler/Twining memo." This memo purports to be written by President Eisenhower's assistant Robert Cutler to General Nathan

F. Twining. It adds authenticity to the MJ-12 documents because it contains a reference to Majestic-12. The memo is dated July 14, 1954 and refers to the scheduling of an MJ-12 briefing. General Twining is MJ-4, according to page two of the documents. The government says that the documents are all fakes.

What concerns us right now are the people who are mentioned as members of MJ-12, because one of them is James Forrestal. The following individuals were described in the documents as "designated members" of Majestic-12 in this order:

Roscoe H. Hillenkoetter (MJ-1)
Vannevar Bush (MJ-2)
James Forrestal (MJ-3)
Nathan F. Twining (MJ-4)
Hoyt Vandenberg (MJ-5)
Detlev Bronk (MJ-6)
Jerome Clarke Hunsaker (MJ-7)
Gordon Gray (MJ-8)
Donald H. Menzel (MJ-9)
Sidney Souers (MJ-10)
Robert M. Montague (MJ-11)
Lloyd Berkner (MJ-12)

The document then noted that General Walter B. Smith replaced James Forrestal on August 1, 1950.

At the time MJ-12 was being formed—1947—President Harry S. Truman appointed Forrestal the first United States Secretary of Defense. During this time Forrestal continued to advocate for complete racial integration of the military services, a policy that was eventually implemented in 1949.

During private cabinet meetings with President Truman in 1946 and 1947, Forrestal had argued against the partition of Palestine to create the new nation of Israel on the grounds it would infuriate Arab countries who supplied oil needed for the US economy and national defense. Instead, Forrestal favored a federalization plan for Palestine.

Because of his inaction, President Truman began receiving threats to cut off campaign contributions from wealthy donors, as well as hate mail, including a letter accusing him of "preferring fascist and Arab elements to the democracy-loving Jewish people of Palestine."

Forrestal was appalled by the intensity and implied threats over the partition question, and he appealed to Truman in two separate cabinet meetings not to base his decision on partition on the basis of political pressure. Forrestal stated to J. Howard McGrath, a Senator from Rhode Island in 1947:

> …no group in this country should be permitted to influence our policy to the point it could endanger our national security.

Forrestal's statement soon earned him the active enmity of some congressmen and supporters of Israel. Forrestal was also an early target of the muckraking columnist and broadcaster Drew Pearson, an opponent of foreign policies hostile to the Soviet Union, who began to regularly call for Forrestal's removal after President Truman named him Secretary of Defense. Pearson, who was known for dirty tactics, once told his own protégé, Jack Anderson, that he believed Forrestal was "the most dangerous man in America" and claimed that if he was not removed from office, he would "cause another world war."

With the wholesale demobilization of most of the US defense force structure, Forrestal resisted President Truman's efforts to substantially reduce defense appropriations. However, he was unable to prevent a steady reduction in defense spending, which resulted in major cuts in defense equipment stockpiles, as well as military readiness.

By 1948, President Harry Truman had approved military budgets billions of dollars below what all the services were requesting which put Forrestal in the middle of a fierce tug-of-war between the President and the Joint Chiefs of Staff. Forrestal continued to be increasingly worried about the Soviet Union as a threat to the United States.

His 18 months as Secretary of the Defense had been a difficult time for the US military establishment: Communist governments came to power in Czechoslovakia and China; the Soviets had imposed a blockade on West Berlin which prompted the US-Berlin Airlift to supply the city; on May 14, 1948 the State of Israel was declared and the Arab–Israeli War followed. Also, negotiations were going on for the formation of NATO.

In 1948 the Governor of New York, Thomas E. Dewey, was expected to win the presidential election. Forrestal had a private meeting with Dewey and it was agreed he would continue as Secretary of Defense under a Dewey administration. Then, just weeks before the election, Forrestal's nemesis, the radio personality and journalist Drew Pearson, published an exposé of the meetings between Dewey and Forrestal.

Truman was angered by Forrestal's maneuvering behind his back and his resistance to the military drawdown that Truman had been ordering. The President abruptly asked the Secretary of Defense to step down. By March 31, 1949, Forrestal was replaced by Louis A. Johnson, a firm supporter of Truman's policies. Around this time, Forrestal had a nervous breakdown and believed that he was being followed and that his phone calls were being monitored.

On the day of Forrestal's resignation from office, March 28, he was reported to have gone into a strange daze and was flown on a Navy airplane to the estate of Undersecretary of State Robert A. Lovett in Hobe Sound, Florida, where Forrestal's wife, Josephine, was vacationing. Dr. William C. Menninger of the Menninger Clinic in Kansas was consulted about Forrestal's strange condition and he diagnosed "severe depression" of the type "seen in operational fatigue during the war."

It was suggested that Forrestal should go to the Menninger Clinic in Kansas, but Forrestal's wife, along with friend Ferdinand Eberstadt and Navy psychiatrist Captain Dr. George N. Raines, decided to send the former Secretary of Defense to the National Naval Medical Center (NNMC) in Bethesda, Maryland.

Forrestal was checked into NNMC five days later. The decision to house him on the 16th floor instead of the first floor was justified because of a need to keep Forrestal's mental health a secret. Forrestal's condition was officially announced as "nervous and physical exhaustion"; his lead doctor, Captain Raines, diagnosed his condition as "depression" or "reactive depression."

Captain Raines reportedly gave Forrestal the following drugs:

1st week: narcosis with sodium amytal.

2nd–5th weeks: a regimen of insulin sub-shock combined with psycho-therapeutic interviews. According to Dr. Raines, the patient overreacted to the insulin much as he had to the amytal and this would occasionally throw him into a confused state with a great deal of agitation and confusion.

4th week: insulin administered only in stimulating doses; 10 units of insulin four times a day, morning, noon, afternoon and evening.

According to Dr. Raines, "We considered electroshock but thought it better to postpone it for another 90 days. In reactive depression if electroshock is used early and the patient is returned to the same situation from which he came there is grave danger of suicide in the immediate period after they return… so strangely enough we left out electroshock to avoid what actually happened anyhow."

What was happening to Forrestal? His life was falling apart. He was fired from his extremely important job. He had delusions and acted in a daze. He thought people were following him and listening to his phone calls. Was he being drugged? Were people actually following him and listening to his phone calls? Had he just lost his marbles, as they say? All of these are possible. One other thing to consider—was Forrestal (MJ-3) still part of the secret group designated MJ-12? It would seem so.

MJ-3 at the National Naval Medical Center

Since entering the NNMC Forrestal gained 12 pounds and seemed to be on the road to recovery. However, in the early morning hours of May 22, 1949, his body, clad only in the bottom half of a pair of pajamas, was found on a third-floor roof below the sixteenth-floor kitchen across the hall from his room.

Forrestal's alleged last written statement, touted in the contemporary press and later by biographers as an implied suicide note, was part of a poem from W. M. Praed's translation of Sophocles' tragedy *Ajax*:

Fair Salamis, the billows' roar,
Wander around thee yet,
And sailors gaze upon thy shore
Firm in the Ocean set.
Thy son is in a foreign clime
Where Ida feeds her countless flocks,
Far from thy dear, remembered rocks,
Worn by the waste of time–
Comfortless, nameless, hopeless save
In the dark prospect of the yawning grave....
Woe to the mother in her close of day,
Woe to her desolate heart and temples gray,
When she shall hear Her loved one's story whispered in her ear!
"Woe, woe!" will be the cry–
No quiet murmur like the tremulous wail
Of the lone bird, the querulous nightingale–

The official Navy review board, which completed hearings on May 31, waited until October 11, 1949 to release only a brief summary of its findings. The announcement, as reported on page 15 of the October 12 *New York Times*, stated only that Forrestal had died from his fall from the window. It did not say what might have caused the fall, nor did it make any mention of a bathrobe sash cord that had been reported by the coroner as being tied around his neck.

A guard had been assigned to watch Forrestal at all times, sitting in the room with the man who held some of the most important secrets in the government. They were on the sixteenth floor of NNMC in a room where the windows had been altered so that they could not open. The main guard assigned to Forrestal was US Navy corpsman Edward Prise. Prise and Forrestal had become good friends and Forrestal even said shortly before his death that he wanted Prise to be his driver when he was eventually released from the hospital.

On the night of Forrestal's death, Prise had been sitting in the room until it became fairly late and Prise's shift ultimately ended. Forrestal told Prise that he didn't need a sleeping pill for that night and was going to stay up and read for a while. Prise was replaced by another Navy corpsman named Robert Wayne Harrison Jr. At some point in his shift Harrison left the room to take his five-minute break. When he returned to the room Harrison said that he was shocked to find Forrestal gone. Racing to find him, Harrison noticed that the cord from Forrestal's robe was tied to a radiator near the open window of the small kitchen across the hall. Looking down from the window he could see Forrestal's dead body 13 floors below him. Had Forrestal tried to hang himself with the cord from his robe as he climbed out of the window—but instead fell to his death—or had he been pushed?

Why had Harrison been able to leave his post, apparently against orders that Forrestal was not to be left alone? Forrestal was due to be released soon said his brother Henry Forrestal shortly afterwards. Henry Forrestal also believed that his brother had been murdered, probably by the Navy.

Incredibly, a full report was not released by the Department of the Navy until April 2004 in response to a Freedom of Information Act request by researcher David Martin. The report said:

> After full and mature deliberation, the board finds as follows:
>
> FINDING OF FACTS
>
> That the body found on the ledge outside of room three eighty-four of building one of the National Naval Medical Center at one-fifty a.m. and pronounced dead at one fifty-five a.m., Sunday, May 22, 1949, was identified as that of the late James V. Forrestal, a patient on the Neuropsychiatric Service of the U. S. Naval Hospital, National Naval Medical Center, Bethesda, Maryland.
>
> That the late James V. Forrestal died on or about May 22, 1949, at the National Naval Medical Center, Bethesda, Maryland, as a result of injuries, multiple, extreme, received incident to a fall from a high point in the tower, building one, National Naval Medical Center, Bethesda, Maryland.
>
> That the behavior of the deceased during the period of his stay in the hospital preceding his death was indicative

> of a mental depression.
>
> That the treatment and precautions in the conduct of the case were in agreement with accepted psychiatric practice and commensurate with the evident status of the patient at all times.
>
> That the death was not caused in any manner by the intent, fault, negligence or inefficiency of any person or persons in the naval service or connected therewith.

With this last statement the US Navy is essentially saying that it investigated itself and found that it was not to blame. Naval Intelligence and other naval groups were not somehow involved. Unfortunately it is difficult to take this statement at face value. One of the difficulties with intelligence services is that it is hard to get accurate statements from them in general, let alone when they undertake investigations in which they themselves are the accused.

According to Nick Redfern in his recent book *Assassinations*,[5] only a few days before Forrestal was sent to Florida and eventually to the hospital, he was visited by his friend Ferdinand Eberstadt. When Eberstadt called to say he wanted to stop by, Forrestal said in a strange voice, "For your own sake, I advise you not to." Eberstadt was a lawyer and a banker who said he was stunned to find all of the curtains shut at Forrestal's home.

In hushed tones Forrestal told him that there were listening devices all over the house. He also said that his life was in danger and that sinister forces were watching his every move. Forrestal then opened one of the blinds and pointed to two men who were standing on a street corner and assured Eberstadt that they were part of the plot and were watching him.

Just then the doorbell rang which threw Forrestal into a panic. A staff member in the house answered the door and spoke with the person there. The staff member then came into the room and told Forrestal that the visitor wanted to speak with him as he was trying to gather support to become the postmaster in his hometown. Could he come in and speak with Forrestal, who might have some advice for him? Forrestal refused to speak with the man and then Eberstadt said that the two watched through an open blind as the man walked directly to the two men on the corner and began speaking with them. This was more evidence of a conspiracy against him, he told Eberstadt grimly.

Eberstadt was never to able figure out if it was just a coincidence that the man talked with the strangers on the corner or not. Perhaps Forrestal was being watched and harassed. Redfern, in his book, suggests that Forrestal was visited by one of the Men in Black who began cropping up around this time. Redfern also thinks that Forrestal was paranoid for good reason and ultimately silenced before he could be let out of the hospital. He was a man who knew too much.

In his 2019 book *The Assassination of James Forrestal*,[3] author David Martin maintains that Forrestal was murdered while at the NNMC. Throughout his book he says that Forrestal was murdered by pro-Israel activists that operated within the government and the US Navy. In light of Forrestal's well-known opposition to the partition of Palestine and the creation of Israel it is easy to see why this may have been a motive for Forrestal's "assassination."

However, his character had already been assassinated by the pro-Israel and pro-Soviet radio personalities Drew Pearson and Walter Winchell and he had been removed from power. Was the pro-Israel lobby, powerful as they were, able to have Forrestal harassed and then assassinated at a Navy hospital? It would seem that Forrestal may have been assassinated, but it is more likely because of his instability and his role in MJ-12 rather than his opposition to Israel.

What happened to James Forrestal? What did he know as a member of MJ-12? Did he know things about Operation Highjump and Antarctica that were too top secret to be discussed with other politicians—something that Forrestal was known to do? Forrestal reportedly saw himself as a "behind the scenes" sort of person, but his importance and stature may have been just too large for even him to control.

Once Forrester was forced out by President Truman did he become a target for mind control and psychological harassment, perhaps with drugs or other methods? Once a person was part of such an insider group as MJ-12, was it really possible to go back to normal civilian life? I think probably not. A combination of drugs and genuine harassment and surveillance would be enough to drive most men to some form of paranoid behavior, much like Forrestal was displaying.

The MJ-12 Documents

But Forrestal and his death aside, during these years the rest of the nation was giddy with excitement that World War II was finally over with the surrender of Japan in the summer of 1945. Says Wikipedia about the celebrations around the world on the surrender:

> On August 6 and 9, 1945, the United States dropped atomic bombs on Hiroshima and Nagasaki, respectively. On August 9, the Soviet Union declared war on Japan. The Japanese government on August 10 communicated its intention to surrender under the terms of the Potsdam Declaration.
>
> The news of the Japanese offer began early celebrations around the world. Allied soldiers in London danced in a conga line on Regent Street. Americans and Frenchmen in Paris paraded on the Champs-Élysées singing "Don't Fence Me In." American soldiers in occupied Berlin shouted "It's over in the Pacific," and hoped that they would now not be transferred there to fight the Japanese. Germans stated that the Japanese were wise enough to—unlike themselves—give up in a hopeless situation, but were grateful that the atomic bomb was not ready in time to be used against them. Moscow newspapers briefly reported on the atomic bombings with no commentary of any kind. While "Russians and foreigners alike could hardly talk about anything else," the Soviet government refused to make any statements on the bombs' implication for politics or science.

But, while the rest of the world could go on with their business (starting new wars perhaps) within two years the top intelligence officials in the US government, guided by Naval Intelligence and directed by Admiral Roscoe H. Hillenkoetter, would face a new challenge—a puzzling onslaught of Unidentified Flying Objects (UFOs). Sometimes flying in a V-formation, these craft were seen flying around various parts of the world, including the United States. Had World War II not ended two years earlier?

Because of this new challenge the secret group called MJ-12 was apparently formed. Also during this time period was the infamous Project Paperclip, the recruitment of former Nazis and German officers, some of them scientists, like Werner von Braun, to come and work in the United States.

When we look at the puzzling MJ-12 documents there seems to be a gnawing feeling that there is something wrong. Operation Majestic-12 appears to be a very real group of twelve senior intelligence officers and scientists within the US government, and all of the people listed in the papers existed.

Some of the names will be instantly familiar to those who study history, while others are more obscure. It is beyond the scope of this book to discuss all of the people on this original MJ-12 list, but each is a fascinating person.

Let me just say that Admiral Roscoe Henry Hillenkoetter (May 8, 1897-June 18, 1982) was clearly the top intelligence officer in the US government immediately after the war. He was the third director of the post–World War II United States Central Intelligence Group (CIG), the third Director of Central Intelligence (DCI), and the first director of the Central Intelligence Agency created by the National Security Act of 1947. He served as DCI and director of the CIG and the CIA from May 1, 1947, to October 7, 1950. He was also the head of MJ-12, according to the document.

Also according to the document, General Walter B. Smith replaced James Forrestal on the MJ-12 panel on Aug. 1, 1950 after Forrestal's death. This makes General Smith a thirteenth member of the original group. A fourteenth name is given to us in the document, that of President Dwight Eisenhower. The document also states that President Truman created Operation MJ-12 by an executive order on September 24, 1947.

But what of the other contents of this document dump? The third page mentions the Kenneth Arnold sighting of boomerang-shaped objects over the Cascade Mountains of Washington State on June 24, 1947. While Arnold described them as boomerang-type

Admiral Roscoe H. Hillenkoetter in 1957.

aircraft, very similar to the German Horton IV flying wing, the MJ-12 document, page three, calls this and the many other sightings "discs." And, indeed, there were many sightings of disks, rather than boomerang-type shapes.

The next paragraph discusses the July 7, 1947 crash of a "disc" near the Air Force base in Roswell, New Mexico, an area that is heavily militarized. The next paragraph mentions the wreckage of a craft and the partially destroyed remains of "four small human-like beings" that had "ejected" from the craft before it exploded and crashed. It also said that the bodies of the four small beings had lain in the desert for approximately a week before they were found and recovered. Yes, one week!

The next page of the MJ-12 documents begins with an assessment that the small saucer craft are scout craft because they have no amenities in them, such as food or a toilet. The document says that Dr. Bronk suggested they call the occupants of the craft EBEs for "Extraterrestrial Biological Entities." The next paragraph says that the occupants of the Roswell craft are thought by Dr. Menzel to be from another solar system.

The third paragraph says that there are symbols in the craft that appear to be writing, and that there is no electrical wiring or other recognizable electronic components. The last sentence says that it was "assumed the propulsion unit was completely destroyed by the explosion which caused the crash."

The next page, page five, starts by saying that there is a need for more information and mentions other secret Air Force projects: Project Sign and Project Grudge. It then mentions another operation that is coordinating these operations. This operation is called Blue Book.

The second paragraph mentions that another UFO crashed on December 6, 1950 on the Mexican side of the Texas-Mexico border near a town called El Indio. The craft was mostly destroyed but what little was recovered was sent to Sandia Air Force base just east of Albuquerque, New Mexico.

The third and final paragraph talks about how we do not know the ultimate intentions of these "visitors" and that we need to keep studying them as well as keep the public calm and not cause a panic. It then mentions a mysterious contingency plan called MJ-1949-04P/78. We never learn what it is, except that it is Top Secret and is attachment "G."

The next pages of the documents are a list of the various attachments, and the final pages are the original memorandum addressed to Forrestal authorizing the creation of MJ-12—spelled out in the document—signed by President Truman and dated September 24, 1947. We also have the supplementary document called the Twining Memo which mentions a briefing by MJ-12.

While the government has denied that the MJ-12 documents are valid, many researchers in the UFO community felt at the time that they were authentic. Curiously, the skeptics were divided in their opinions on who had created the fake documents. Some pointed to Moore and Shandera as the creators of the supposed hoax. But, other skeptics felt that the forgery had occurred—get this—within the US government itself in order to plant disinformation with Shandera and his colleagues. That the CIA or Air Force Intelligence, or some other intelligence agency, had created these fake documents about MJ-12 in order to spread disinformation is interesting.

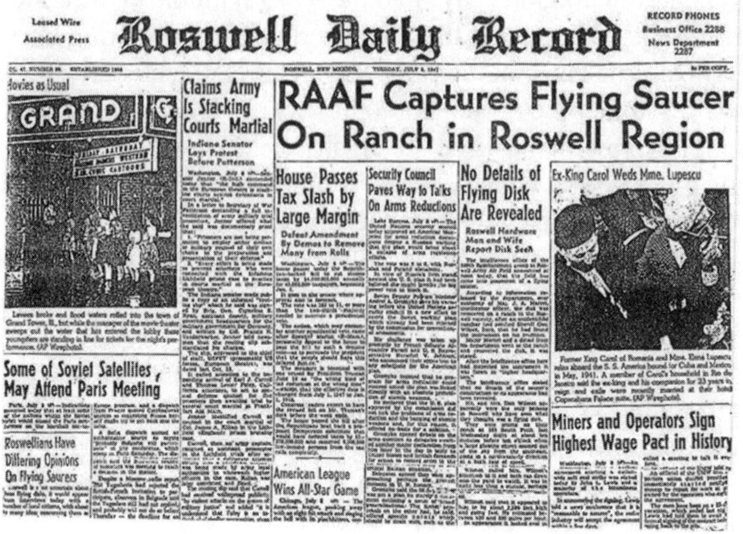

Leased Wire
Associated Press

Roswell Daily Record

RECORD PHONES
Business Office 2288
News Department 2287

RAAF Captures Flying Saucer On Ranch in Roswell Region

Movies as Usual

Claims Army Is Stacking Courts Martial

Indiana Senator Lays Protest Before Patterson

House Passes Tax Slash by Large Margin

Defeat Amendment By Demos to Remove Many From Rolls

Security Council Paves Way to Talks On Arms Reductions

No Details of Flying Disk Are Revealed

Roswell Hardware Man and Wife Report Disk Seen

Ex-King Carol Weds Mme. Lupescu

Some of Soviet Satellites May Attend Paris Meeting

Roswellians Have Differing Opinions On Flying Saucers

American League Wins All-Star Game

Miners and Operators Sign Highest Wage Pact in History

The *Roswell Daily Record* dated July 8, 1947.

This "disinformation" was handed anonymously to several of the top UFO researchers in the 1980s. But why would there be disinformation like this? Most instances of disinformation involve some real information. What was trying to be hidden with fake documents about MJ-12? What was the real information? Was it to fool gullible UFO researchers that there was once a secret MJ-12 group within the government and intelligence agencies that never existed? Would four dead and decomposing extraterrestrials really be lying out in the desert for a week after a UFO crash only miles from a major Air Force base?

There is nothing really classified about the discussion of the Kenneth Arnold sighting or the Roswell crash. All of this was largely "public knowledge" and there is not much real meat in this Top Secret briefing, except for the creation of MJ-12 itself. This group and its members is the real story, whatever their activities may have been.

Is it possible that what was really discussed by MJ-12 were the results of the invasion of Antarctica the year before? With all of the unusual events happening at the end of World War II, including UFO sightings and a large number of missing submarines, it would seem that the most pressing issues were still-existing Nazi bases in various parts of the world, as well as the rise of the Soviet Union and other events in China, Korea and Vietnam.

It would seem that the MJ-12 documents were hiding a larger threat than small human-like EBEs. And if the claims about extraterrestrials in the MJ-12 documents are genuine, did the group conclude that they were coming from Antarctica? Operation Highjump, to be discussed in the next chapter, ended in late February of 1947 and the MJ-12 group was formed in September later that year.

If those participating in Operation Highjump witnessed flying saucers in Antarctica, as has been reported, then perhaps MJ-12 was formed as a think tank to be the "spin doctors" on the strange events happening around the world and the flying saucer craze that swept through Europe, Asia, the Americas and elsewhere. Where were these craft coming from—outer space? Or, were they coming from Antarctica? Was this why Forrestal was drugged, hospitalized, and murdered? He was too vocal in his discussions with other people, including Senator McCarthy—famously so. Can this be the reason he ultimately ended up in a Navy hospital being given all sorts of mind control drugs, including scopolamine?

Furthermore, Forrestal's anti-Soviet stance was causing trouble with the prevailing idea of having some cooperation with the Russians while maintaining an adversarial position on the world stage. Ultimately the Cold War was one of cooperation, as were the space initiatives—the "space race"—culminating with the International Space Station and the dissolution of the Communist Soviet Union in 1989. How can you have a war with someone and share a space station with that entity at the same time? Ask NASA.

Something was happening in Antarctica and it didn't have to do with the Russians. They had plenty of frozen territory in the north to deal with. Operation Highjump had returned from Antarctica and had shaken the Navy with what they had discovered. James Forrestal knew too much about this activity and was considered a loose cannon on deck. He had to be eliminated by first sending him to the Navy loony bin and then having him commit suicide. MJ-12, however, would continue on without him.

So, as we end this chapter we need to ask, what happened in Antarctica during Operation Highjump?

TOP SECRET / MAJIC
EYES ONLY 002

* TOP SECRET *

EYES ONLY COPY ONE OF ONE.

SUBJECT: OPERATION MAJESTIC-12 PRELIMINARY BRIEFING FOR PRESIDENT-ELECT EISENHOWER.

DOCUMENT PREPARED 18 NOVEMBER, 1952.

BRIEFING OFFICER: ADM. ROSCOE H. HILLENKOETTER (MJ-1)

NOTE: This document has been prepared as a preliminary briefing only. It should be regarded as introductory to a full operations briefing intended to follow.

* * * * * *

OPERATION MAJESTIC-12 is a TOP SECRET Research and Development/ Intelligence operation responsible directly and only to the President of the United States. Operations of the project are carried out under control of the Majestic-12 (Majic-12) Group which was established by special classified executive order of President Truman on 24 September, 1947, upon recommendation by Dr. Vannevar Bush and Secretary James Forrestal. (See Attachment "A".) Members of the Majestic-12 Group were designated as follows:

Adm. Roscoe H. Hillenkoetter
Dr. Vannevar Bush
Secy. James V. Forrestal*
Gen. Nathan F. Twining
Gen. Hoyt S. Vandenberg
Dr. Detlev Bronk
Dr. Jerome Hunsaker
Mr. Sidney W. Souers
Mr. Gordon Gray
Dr. Donald Menzel
Gen. Robert M. Montague
Dr. Lloyd V. Berkner

The death of Secretary Forrestal on 22 May, 1949, created a vacancy which remained unfilled until 01 August, 1950, upon which date Gen. Walter B. Smith was designated as permanent replacement.

* TOP SECRET *

TOP SECRET / MAJIC
EYES ONLY EYES ONLY T52-EXEMPT (E)
002

A page from the MJ-12 documents dated November 18,1952.

A-3

TOP SECRET / MAJIC

EYES ONLY

003

* TOP SECRET *

EYES ONLY

COPY ONE OF ONE.

On 24 June, 1947, a civilian pilot flying over the Cascade Mountains in the State of Washington observed nine flying disc-shaped aircraft traveling in formation at a high rate of speed. Although this was not the first known sighting of such objects, it was the first to gain widespread attention in the public media. Hundreds of reports of sightings of similar objects followed. Many of these came from highly credible military and civilian sources. These reports resulted in independent efforts by several different elements of the military to ascertain the nature and purpose of these objects in the interests of national defense. A number of witnesses were interviewed and there were several unsuccessful attempts to utilize aircraft in efforts to pursue reported discs in flight. Public reaction bordered on near hysteria at times.

In spite of these efforts, little of substance was learned about the objects until a local rancher reported that one had crashed in a remote region of New Mexico located approximately seventy-five miles northwest of Roswell Army Air Base (now Walker Field).

On 07 July, 1947, a secret operation was begun to assure recovery of the wreckage of this object for scientific study. During the course of this operation, aerial reconnaissance discovered that four small human-like beings had apparently ejected from the craft at some point before it exploded. These had fallen to earth about two miles east of the wreckage site. All four were dead and badly decomposed due to action by predators and exposure to the elements during the approximately one week time period which had elapsed before their discovery. A special scientific team took charge of removing these bodies for study. (See Attachment "C".) The wreckage of the craft was also removed to several different locations. (See Attachment "B".) Civilian and military witnesses in the area were debriefed, and news reporters were given the effective cover story that the object had been a misguided weather research balloon.

* TOP SECRET *

EYES ONLY TOP SECRET / MAJIC

EYES ONLY

T52-EXEMPT (E)

003

A page from the MJ-12 documents dated November 18,1952.

A-4
TOP SECRET / MAJIC
EYES ONLY
004

* TOP SECRET *

EYES ONLY

COPY ONE OF ONE.

A covert analytical effort organized by Gen. Twining and Dr. Bush acting on the direct orders of the President, resulted in a preliminary concensus (19 September, 1947) that the disc was most likely a short range reconnaissance craft. This conclusion was based for the most part on the craft's size and the apparent lack of any identifiable provisioning. (See Attachment "D".) A similar analysis of the four dead occupants was arranged by Dr. Bronk. It was the tentative conclusion of this group (30 November, 1947) that although these creatures are human-like in appearance, the biological and evolutionary processes responsible for their development has apparently been quite different from those observed or postulated in homo-sapiens. Dr. Bronk's team has suggested the term "Extra-terrestrial Biological Entities", or "EBEs", be adopted as the standard term of reference for these creatures until such time as a more definitive designation can be agreed upon.

Since it is virtually certain that these craft do not originate in any country on earth, considerable speculation has centered around what their point of origin might be and how they get here. Mars was and remains a possibility, although some scientists, most notably Dr. Menzel, consider it more likely that we are dealing with beings from another solar system entirely.

Numerous examples of what appear to be a form of writing were found in the wreckage. Efforts to decipher these have remained largely unsuccessful. (See Attachment "E".) Equally unsuccessful have been efforts to determine the method of propulsion or the nature or method of transmission of the power source involved. Research along these lines has been complicated by the complete absence of identifiable wings, propellers, jets, or other conventional methods of propulsion and guidance, as well as a total lack of metallic wiring, vacuum tubes, or similar recognizable electronic components. (See Attachment "F".) It is assumed that the propulsion unit was completely destroyed by the explosion which caused the crash.

* TOP SECRET *

EYES ONLY TOP SECRET / MAJIC T52-EXEMPT (E)
EYES ONLY
004

A page from the MJ-12 documents dated November 18,1952.

A-5

TOP SECRET / MAJIC
EYES ONLY 005

* TOP SECRET *

EYES ONLY COPY ONE OF ONE.

A need for as much additional information as possible about these craft, their performance characteristics and their purpose led to the undertaking known as U.S. Air Force Project SIGN in December, 1947. In order to preserve security, liason between SIGN and Majestic-12 was limited to two individuals within the Intelligence Division of Air Materiel Command whose role was to pass along certain types of information through channels. SIGN evolved into Project GRUDGE in December, 1948. The operation is currently being conducted under the code name BLUE BOOK, with liason maintained through the Air Force officer who is head of the project.

On 06 December, 1950, a second object, probably of similar origin, impacted the earth at high speed in the El Indio - Guerrero area of the Texas - Mexican boder after following a long trajectory through the atmosphere. By the time a search team arrived, what remained of the object had been almost totally incinerated. Such material as could be recovered was transported to the A.E.C. facility at Sandia, New Mexico, for study.

Implications for the National Security are of continuing importance in that the motives and ultimate intentions of these visitors remain completely unknown. In addition, a significant upsurge in the surveillance activity of these craft beginning in May and continuing through the autumn of this year has caused considerable concern that new developments may be imminent. It is for these reasons, as well as the obvious international and technological considerations and the ultimate need to avoid a public panic at all costs, that the Majestic-12 Group remains of the unanimous opinion that imposition of the strictest security precautions should continue without interruption into the new administration. At the same time, contingency plan MJ-1949-04P/78 (Top Secret - Eyes Only) should be held in continued readiness should the need to make a public announcement present itself. (See Attachment "G".)

* TOP SECRET *

EYES ONLY TOP SECRET / MAJIC
EYES ONLY T52-EXEMPT (E)

A page from the MJ-12 documents dated November 18,1952.

A-9

Notes by S.T. Friedman: This document was found after a few days of searching in the just declassified boxes of Record Group 341 in Mid 1985 by Jaime Shandera and William Moore. Stanton Friedman had discovered during a visit to the National Archives in March 1985 that the RG was in the process of being classification reviewed. Post cards were received hinting that checking the file would be a good idea. This memo clearly has nothing to do with anything else in Box 189 where it was found. Most likely it was planted there during the classification review which involved many teams of 4 each working for a few weeks in a location where they were able to bring in notes, files, brief cases etc. The item in its original form is a carbon on Dictation Onion Skin by Fox Paper. It is discolored around the edges. My best bet for the actual author is James S. Lay who was Exec. Sec of NSC and worked very very closely with Cutler and met "off the Record" with Ike at the WH on July 14, 1954. The mark through the classification is red STF →

July 14, 1954

~~TOP SECRET RESTRICTED SECURITY INFORMATION~~

MEMORANDUM FOR GENERAL TWINING

SUBJECT: NSC/MJ-12 Special Studies Project

The President has decided that the MJ-12 SSP briefing should take place during the already scheduled White House meeting of July 16, rather than following it as previously intended. More precise arrangements will be explained to you upon arrival. Please alter your plans accordingly.

Your concurrence in the above change of arrangements is assumed.

Note that the last sentence is almost identical to the wording of another TS Cutler-Twining memo found at the Library of Congress in the Twining papers. STF

ROBERT CUTLER
Special Assistant
to the President

Note that there is no signature and no /s/
STF

Authority NND 857013
By STF Date 1/12/87

COPY
from
[TH]E NATIONAL ARCHIVES
[Rec]ord Group No. RG 341, Records of the Headquarters United States Air Force

The so-called Twining Memo mentions the MJ-12 group in 1954 indicating that it did exist. However, for some reason the original documents contained disinformation.

Chapter Three

A Battle in Antarctica

I'd rather have questions that can't be answered
than answers that can't be questioned.
—Richard P. Feynman

As mentioned in the last chapter, one of the last significant operations that Forrestal authorized after the end of WWII was Operation Highjump, a costly invasion of Antarctica led by the famous Navy admiral and explorer Richard E. Byrd. Many researchers believe that the final battles of World War II were fought in Antarctica. This is based on the theory that the German's had created a large, secret base in Queen Maud Land which they continued to use even after Germany's official surrender on May 8, 1945.

This secret base was set up in January-February of 1939 when a secret German expedition visited Antarctica and declared this part of the continent to be part of the Third Reich and renamed it Neuschwabenland. As mentioned in chapter one, it is thought that Rudolf Hess was a member of this secret expedition. Just prior to WWII it is thought that the Germans were preparing a number of secret bases in remote corners of the world, including Antarctica, Greenland, the Canary Islands, the former Spanish colony south of Morocco known as Spanish Sahara, and even in Tibet (to be discussed in a later chapter). During the war they established other secret bases in occupied countries such as Norway, Poland and Ukraine.

Because of this suspected German base in Neuschwabenland, between 1943 and 1945 the British launched their own secret wartime Antarctic operation that was code named Tabarin. Men from the Special Air Services Regiment (SAS), Britain's covert forces for operating behind the lines, appeared to be involved. This was followed up with Operation Highjump in 1946-1947 and Operation Windmill the next year. Both were secret, classified operations when they occurred. Finally there was another classified US operation in 1958 called Operation Argus. During Operation Argus three nuclear weapons were detonated in the vicinity of Antarctica. No one is quite sure what happened during Operation Argus but some researchers think that the atomic detonations were meant to destroy some of the Nazi facilities that were still operating in Antarctica.

Let us first look at the secret British military action in Antarctica code named Tabarin.

Tabarin: Britain's Secret Antarctic Operation

Operation Tabarin was a secret British Antarctic expedition that lasted from 1943 to 1945. It was launched under the pretense of patrolling the Antarctic for German commerce raiders and U-boats that threatened Allied shipping during the course of World War II. The expedition resulted in the establishment of a number of bases, thus strengthening British territorial claims to the area. Because of the British presence in the Falkland Islands, which are fairly close to Antarctica, a secret British base in Antarctica was quite feasible for the British military. While parts of the operation were declassified in the 1950s it is probable that aspects of Operation Tabarin are still military secrets today. The British, as well as other governments, are famous for keeping large amounts of information about WWII still classified today. For instance, we don't know if any British submarines ever visited any of the Antarctic bases established during Operation Tabarin. We do not know if there was ever any conflict with German forces in Antarctica or if German ships were ever sunk in the vicinity.

One theory has it that British SAS commandos attacked the German base in Neuschwabenland and were defeated by the larger force of Germans. Actual submarine pens, protected inside a granite cave, are surmised to have existed in Neuschwabenland.

Fanciful battles between the British and Germans in Antarctica during the war may be one thing, but the official account of the operation paints a picture of a terrifically boring time in Antarctica where they collected some rocks and fossils, and took meteorological and topographical data, and glaciological measurements. The operation was terminated January 14, 1946 when all the participants returned to the Falkland Islands and then to Britain.

With the outbreak of World War II, Allied shipping across the globe became vulnerable to attacks by German Navy U-boats and commerce raiders. The British Falkland Islands and the waters of the Antarctic were no exception and it was known that a number of U-boats and raiders were in the South Atlantic.

One such commerce raider was the *Pinguin,* a German auxiliary cruiser that was known to the German naval Kriegsmarine as *Schiff 33* and known to the British Royal Navy as *Raider F*. The *Pinguin* was the most successful commerce raider of WWII and became something of a legend.

Formerly a freighter named *Kandelfels*, she was built in 1936, and was owned and operated by the Hansa Line in Bremen, Germany. She was the sister-ship of the commerce raider *Kybfels*, and a half-sister of a freighter called the *Goldenfels* which was converted into a commerce raider renamed *Atlantis*. The choosing of such a name shows the influence of the Thule Society, even in the realm of these converted commercial raiders.

These former freighters still had their large cargo holds for goods—goods that would largely be stolen from other ships. But they were now heavily armed with torpedoes, canons, anti-aircraft guns and even seaplanes. Their missions were to raid Allied vessels and engage with unarmed cargo ships and confiscate their cargo. They ranged throughout the Atlantic and Indian Ocean during WWII.

In the winter of 1939–1940, the *Pinguin* was requisitioned by the Kriegsmarine and converted to a warship in Bremen. Her main armament, six 150 mm guns, were taken from the obsolete battleship *Schlesien,* and she was also fitted with one captured French 75 mm cannon, one twin 37 mm anti-aircraft mounting, four 20 mm anti-aircraft guns, and two single 53.3 cm torpedo tubes with 16 torpedoes. She was supplied with two Heinkel He 114A-2 seaplanes and 300 sea mines. The ship also carried 25 G7a torpedoes and 80 U-boat mines for replenishing U-boats at sea. Some of these mines and torpedoes may have gone to the German base in Neuschwabenland where U-boats could be replenished with

The German Raider named *Pinguin* in 1941.

cargo brought by commerce raiders operating in the South Atlantic.

After raiding in the Indian Ocean, on January 13, 1941, the *Pinguin* began terrorizing the unarmed and unescorted Norwegian whaling fleet which had relocated to Antarctic waters because they had depleted the whale population of the Northern Hemisphere through overexploitation. The *Pinguin* managed to seize a haul of 20,320 tons of whale oil from the whalers *Solglimt*, *Pelagos* and *Ole Wegger*. This was ultimately one of the largest prizes seized by a commerce raider during the war. Shortly afterwards, the Norwegian whaler *Thorshammer* alerted the British authorities on the Falklands, and the armed merchant cruiser *Queen of Bermuda* began patrolling the area between South Georgia, the South Shetland Islands and the Weddell Sea of Antarctica.

The *Pinguin* then headed towards the shipping routes between the Persian Gulf and Mozambique. The *Pinguin* began searching for a tanker to loot near the entrance of the Persian Gulf. On May 7 a small tanker was spotted, the British tanker *British Emperor*. The *Pinguin* signaled to the tanker to heave to, but the British ship refused to obey and began transmitting distress signals describing her attacker and identifying herself as the British tanker *British Emperor*. *Pinguin's* gunners fired a salvo of deliberate near misses to encourage *British Emperor* to stop, and then fired a salvo that destroyed the tanker's bridge and wheelhouse.

But before the tanker *British Emperor* could be captured the British destroyer *Cornwall* approached the *Pinguin* from the south. The *Pinguin* had been disguised as a British ship and was flying a British flag. It now dropped its disguise and ran up the German battle flag, turned full broadside to the British destroyer and opened up with five guns simultaneously. The *Cornwall* suffered a failure in the electrical circuit that controlled the training of her main gun turrets.

The *Cornwall* broke off and retired out of range of *Pinguin*'s guns to carry out repairs. The *Cornwall* attempted to launch a waiting Walrus seaplane to bomb the *Pinguin*, however, it had suffered damage and was unable to take off. Out of range of *Pinguin*'s guns, the crew repaired the damage to the turret circuits on the *Cornwall*. They then pursued the *Pinguin,* and once it was spotted the *Cornwall*'s gunners soon began to fire shells at the raider, the first hit bringing down the foremast. The German commander instantly gave orders to release all the prisoners and to set the scuttling charges and abandon the ship.

At that very moment a four-gun salvo from the *Cornwall*'s 8-inch forward turrets destroyed *Pinguin*. The first shell struck the foredeck wiping out the two 150 mm guns on the forecastle head and their crews. The second shell hit the meteorological office and shattered the bridge killing nearly everyone instantly. The third shell devastated the engine room. The fourth shell exploded in Hold Number 5 detonating the 130 high-explosive mines stored there, ripping the *Pinguin* to pieces. Flames shot thousands of feet into the air and fragments of *Pinguin* were scattered across the surface of the sea. The German raider *Pinguin* was gone within only five seconds.

The action lasted just 27 minutes from beginning to end. The *Cornwall*'s boats picked up 60 members of *Pinguin*'s crew and 24 of her former prisoners. Of the 401 Germans aboard *Pinguin* only three officers and 57 petty officers and men survived. Of the 238 prisoners on the *Pinguin* only 24 survived.

While the *Cornwall* had seen fierce action and sunk the *Pinguin*, back near the Falklands the *Queen of Bermuda* was still on patrol. On March 5, the *Queen of Bermuda* visited Deception Island, one of several islands just off the Antarctic coast, where the British crew set fire to a coal dump and sabotaged oil tanks belonging to an abandoned Norwegian whaling station in order to prevent their use as a supply base by the Germans. Little did they know that the Germans had a much larger base to the east. All of these missions and operations had been top secret and the British had no knowledge of the 1938-1939 German expedition to Antarctica at this time, except for intelligence reports.

Also, the British perceived the entry of Japan as another Axis power in the war as a threat to their overseas possessions. Surprisingly, they were afraid that the Japanese would attack the Falklands and overwhelm these sparsely populated islands near the southern tip of South America.

The Falklands were protected by an insufficient force of 330 local volunteers who had few weapons, making an amphibious Japanese invasion quite possible. The British reasoned that the Japanese could overwhelm the islands and locate thousands of soldiers there to supply Japanese subs and destroyers that were keeping supply ships from going from the Atlantic around Cape Horn and into the Pacific, cutting off supplies to Australia and India. The Japanese could also use the Falklands to harass shipping that was going around the Cape of Good Hope in South Africa to India and the Middle East.

The war also threatened to reignite the longstanding Falkland Islands sovereignty dispute with Argentina, a country that was officially neutral during the war but was known to side with Germany, a country with a lot of financial interest in Argentina. In January of 1942, Argentina's Comisión Nacional del Antártico dispatched a ship to Deception Island off the Antarctic coast; the ship then visited the Melchior Islands, Palmer Archipelago and Winter Island. Argentine flags were raised in all of these locations and all territories south of 60° S and between 25° W and 68.34° W were declared annexed and part of Argentina. Clearly, the secret war in Antarctica had begun.

On January 28, 1943, the British Colonial Office dispatched the armed merchant cruiser HMS *Carnarvon Castle* to Deception Island after it had falsely claimed that a German commerce raider had been spotted in the area. Upon reaching Deception Island, *Carnarvon*'s crew replaced the Argentinian flag with the Union Jack and placed

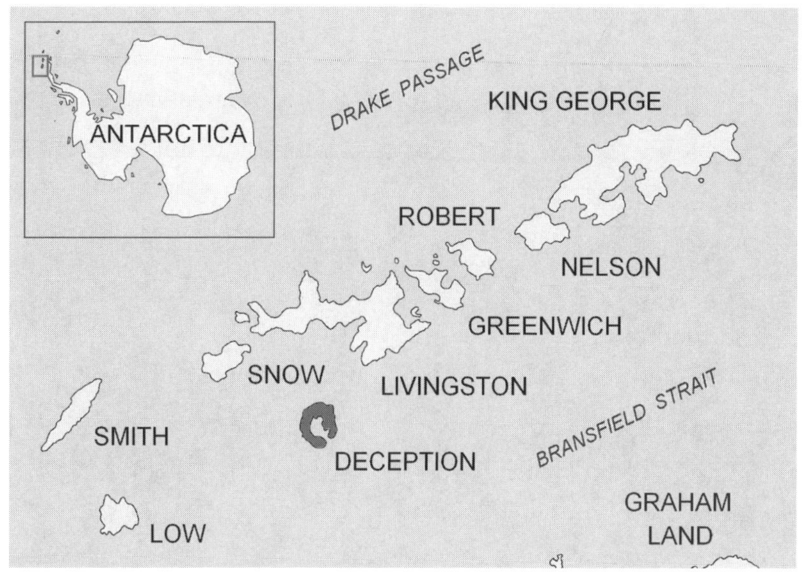

A map showing Deception Island in the South Shetland Islands.

four British Crown Land signs.

In May of 1943 Operation Tabarin began from the Falkland Islands. It was originally named Operation Bransfield after the Antarctic explorer Edward Bransfield, but was quickly changed to Operation Tabarin (after a famous Paris cabaret called Bal Tabarin) so as not to reveal its objectives to anyone not part of the mission, including the Americans and other British officers. The name Tabarin would suggest that the operation had something to do with France or Paris which was completely wrong.

British polar experts Neil Mackintosh, James Wordie and Brian Roberts took up the planning of the endeavor. Part of the crew were special forces SAS officers who were well armed. Experienced Scottish marine biologist and polar explorer Lieutenant James Marr was selected as the head of the expedition. Marr, now a Lieutenant Commander, flew to Iceland where he procured the Norwegian sealer *Veslekari* which was assigned the name HMS *Bransfield*. He then proceeded to the Falkland Islands.

The *Bransfield* then went to Port Lockroy, a natural harbor on Wiencke Island, a large island in the Palmer Archipelago. The Palmer Archipelago is just off the Antarctic coast of the peninsula that juts up to the southern tip of South America.

The cargo was unloaded at the site of the newly established

A photo of Operation Tabarin at Deception Island in the South Shetland Islands.

Station A on Goudier Island in the Port Lockroy harbor. The first stage of the construction of the main hut christened Bransfield House was completed on 17 February.

On March 23, 1943 the secret Port Lockroy Post Office began its operation, accepting mail from and delivering letters back to Britain via the relatively frequent trips to Port Stanley in the Falklands. Incredibly this post office is still in use today and is a popular stop for tourist cruises to Antarctica where the passengers get their passports stamped and can buy stamps and postcards to send to friends from what is thought to be the loneliest post office in the world.

In April, the Stanley Post Office in the Falklands burnt to the ground and was not functioning for many months. Therefore, correspondence was now passing through Montevideo, Uruguay—a neutral country—and thus the expedition's existence came to be known to the outside world.

However, according to the reports that came out in the 1950s, not a whole lot happened at Port Lockroy and those stationed there did their best to avoid cabin fever. They took rock samples and collected lichen samples and did some experiments concerning the rate of ice melting and other things. Engaging with a secret German submarine base was not one of the listed activities that you can find about Operation Tabarin. No daring assault on a rock cavern with submarine pens. No, they were bored out of their minds collecting lichen. Unless you have a lot exciting things to do, life can get pretty boring in Antarctica.

On February 3, 1945 the 550-ton sealer *Eagle* arrived at Port Lockroy with new men and equipment, and throughout the spring they continued to build more buildings, including some on other nearby islands, to strengthen Britain's claim to this bit of Antarctic territory. With the end of WWII the men ultimately returned to Port Stanley in the Falklands and then back to Britain, including the SAS troops. The station at Port Lockroy continued to be manned by British personnel but officially there still was not a lot of activity.

The end of World War II led to renewed interest in the Antarctic region. The United States refused to recognize any foreign territorial claims to Antarctica. Argentina and Chile signed the Argentine-Chilean Agreement on Joint Defense of "Antarctic Rights," a defense agreement that envisioned potential military action over disputed Antarctic lands.

Chile organized its First Chilean Antarctic Expedition in 1947–1948. The Chileans built several bases on the continent and Chilean President Gabriel González Videla went to inaugurate one of its bases personally, thereby becoming the first head of state to set foot in Antarctica—except maybe Adolf Hitler, depending on which tales you may believe.

Britain, on the other hand, continued the operation of the bases built during Operation Tabarin by transferring them to the newly-established Falkland Islands Dependencies Survey.

The Port Lockroy base continued to operate as a British research station until January 16, 1962. It was abandoned for some years but in 1996, the Port Lockroy base was renovated and is now a museum operated by the United Kingdom Antarctic Heritage Trust. Participants of Operation Tabarin were awarded the British Polar Medal in 1953.

As we have noted, today, Port Lockroy is one of the most popular tourist destinations for cruise ship passengers in Antarctica. Proceeds from the small souvenir shop at the museum fund the maintenance of the site and other historic sites and monuments in Antarctica. The Trust collects data for the British Antarctic Survey to observe the effect of tourism on penguins. Half the island is open to tourists, while the other half is reserved for penguins. A staff of four at the post office typically processes 70,000 pieces of mail sent by 18,000 visitors that arrive during the five-month Antarctic cruise season.

Operation Tabarin, at the very least, proves that major powers had a great deal of interest in Antarctica during the war. Both Germany and Britain are now known to have established secret bases in Antarctica during the war, with the British base—once secret, now overt—occupied until 1962. One has to wonder, did the Americans create a secret base in Antarctica during this period as well? Officially, they did not.

A Secret Battle Between the British and Neuschwabenland?

Before we leave the subject of Operation Tabarin, we must mention a series of three articles that appeared in *Nexus* magazine starting in October of 2005. *Nexus* magazine is an Australian periodical that is published in the United States and Britain as well, and has a large readership in the UK, a country that loves magazines, even today.

The article (parts one, two and three) were entitled "Britain's Secret War in Antarctica" and the author was said to be James Robert. In a biography at the end of the three parts we are told that James Robert is a civil servant for the UK Ministry of Defense and a World War II historian and writer. His family has a military background. The biography says that Robert has spent considerable time and gone to some expense to travel, researching his history topics. The biography promises us a book and lists an email contact address that stopped functioning almost immediately after the publication of the three parts of the article.

The central character and informant in the article is an unnamed SAS soldier who was active during the Second World War and whom James Robert interviewed twice over a ten-year span of time, he claims.

The article starts out with a brief outline of Operation Tabarin, saying that it was a top secret operation to counter suspected German activity in Antarctica and was kept secret from the Americans, French and Russians during the war. Robert's informant says the British also wanted a propaganda victory if, in fact, the Nazis were in Antarctica so they could then claim to have won the last battle of World War Two all by themselves.

According to Robert's informant, he was not sent on the first mission to Neuschwabenland. This first mission had failed. The SAS group deployed there sent out garbled radio transmissions indicating problems, but confirming that there was indeed a German presence complete with an access tunnel leading to the Nazi stronghold within a mountain near the coast. The informant says that only two survivors escaped back out of the Nazi base. A final message was sent in July of 1945 (two months after the official surrender of Germany), that said in fear, "the Polar Men have found us."

The informant allegedly told James Robert that he and his SAS team were trained in polar survival and then parachuted to within striking range of the German base from Port Lockroy. They gained access to the large German base through a long, man-made tunnel and found a huge cavern, artificially illuminated with docks for U-boats and hangars for some sort of high tech aircraft. The facility was powered geothermally with a secondary power source of unknown origin. The mission of the SAS team was to blow this facility up.

The *Nexus* magazine article starts to get even stranger here because the informant now tells James Robert that there were Nazi guards in uniform at this huge underground base, but also guards that were hairy humanoids, like bigfoot or yetis, that the SAS called "the Polar Men." It is explained that these Polar Men were of unknown origin although it is said they were the products of Nazi science. Completely covered in hair, they possessed resistance to the intense cold. These hairy apemen were said to eat human flesh.

The remaining portion of the tale is about the attempt to blow up parts of the facility and then escape the underground base that was equipped with submarine pens and "strange aircraft," presumably flying saucers. Henry Stevens, who recounts and comments on the article in *Dark Star*,[46] remarks that the cavern may have been a volcanic bubble but the geothermal activity is certainly associated with a fault line. He says the Germans had a base in Greenland where there was a coal deposit and speculates that a coal deposit may have been found in Antarctica, as the continent does have them. Stevens suggests that without a local energy source already being in place, it is doubtful that any long-term presence could have been maintained.

The brief description of the base may reflect an actual German base in Neuschwabenland, but what about the element of the Polar Men? This strange detail seems to cancel any validity to the story. Says Stevens:

> What is unnecessary to this tale are the Polar Men. The Polar Men actually add nothing to the facts presented in this story. Since they add nothing to the fact-pattern and serve as a gateway to attack the story and the story certainly could be told without them, only two reasons present themselves. First, they are added to spice up the story, a sort of Guns of Navarone meets the Abominable Snowman. As such a ploy it certainly succeeds! But the second alternative is that this was exactly the way it was described to Mr. Robert. In other words, it is the truth regardless of how it sounds.
>
> *Nexus* magazine does not print fictional stories so they must have done some checking on Mr. Robert. Assuming they did and found his biography or at least the more verifiable portions, like his employment at the UK Ministry of Defense, checked out, we are left with the problem that this portion of the tale may be true, that the Germans enlisted the assistance of an unknown race of hairy, cold-tolerant men who functioned as security guards for them in this environment. According to the tale, the British did manage to kill one Polar Man and he was a man, only covered in fur growing right out of his skin. But, because of the tale's internal weakness involving the Polar Men and without positive proof for their existence, this story is improbable and that is that.[46]

So what are we to make of the story of a battle between British SAS soldiers and Nazis inside the secret base in Neuschwabenland? As we will see in the coming chapters, such a base is almost certain to have existed. Did the British suspect a German base in Antarctica? They probably did. Did SAS troops try to attack and destroy the base? They may have. Were there hairy Polar Men, genetically engineered by Nazi scientists, guarding the secret base?

Probably not. The *Nexus* magazine article seems to be fiction, but fiction based on intriguing fact. Was James Robert's informant a real person, and if so, was he really a former SAS officer? The British are fond of hoaxes and pranks—was this story just a hoax? There is often a reason for disinformation, and most disinformation has some truth within it, while other parts are completely false. As far as the British SAS having ventured inside the Neuschwabenland base in July of 1945, we may never know the truth as the government of the UK (and the Americans and Russians) are well known to keep much of what happened during WWII classified.

The Strange Story of Operation Highjump

The first Americans to work in the Antarctic were sealers and whalers who discovered many sub-Antarctic islands. They were some of the first to explore parts of the great peninsula jutting out of the Antarctic mainland toward South America. Among them was Nathaniel Palmer, who was among the first mariners to see Antarctica, while on board the USS *Hero* in 1820. James Eights, a geologist from Albany, New York, became the first American scientist to work in Antarctica. In 1830, from the ship *Annawan*, Eights made voyages in the South Shetland Islands and westward along the Antarctic Peninsula, mapping the area as best he could.

Expeditions sponsored by several nations approached the Antarctic continent early in the 19th century. Among the leaders was US Navy lieutenant Charles Wilkes who commanded an expedition in 1839–1840 that was the first to prove the existence of the icy continent. His expedition mapped about 1,500 miles (2,400 km) of the coastline in the Australian and Indian quadrants.

Just prior to WWII, in 1928–1930 and again in 1933–1935, Admiral Richard E. Byrd led two privately sponsored expeditions, one that included the first flight over the South Pole in 1929. This sparked US interest in Antarctica and that led to the US Antarctic Service Expedition from 1939 to 1940, under the leadership of the US Navy, that maintained temporary bases at Marguerite Bay and the Bay of Whales, which is technically a natural ice harbor on the Ross Ice Shelf and not on actual land.

America's next incursion to Antarctica would be Operation Highjump, the largest single expedition ever to explore Antarctica.

Admiral Byrd in Antarctica in 1939.

Approved in the summer of 1946 by Secretary of the Navy James Forrestal, the carrier group of the naval task force set sail from Norfolk, Virginia and was led by Admiral Richard E. Byrd. It involved 4,700 men, 13 ships (including submarines and an aircraft carrier) and multiple aircraft. The fleet was designated Task Force 68 with Rear Admiral Richard H. Cruzen, USN, as the commanding officer (after Byrd).

Operation Highjump was launched on August 26, 1946 and was supposed to last for one year, until August of 1947, but was abruptly terminated in late February 1947, six months earlier than planned.

According to the government, Operation Highjump's primary mission was to establish an Antarctic research base, to be called Little America IV. Conspiracy theories abound as to Operation Highjump's ultimate mission in Antarctica and why it returned six months early. Says a curious entry in *Wikipedia*:

In 1939 Admiral Byrd took this Snow Cruiser to Antarctica.

Operation Highjump has become a topic among UFO conspiracy theorists, who claim it was a covert U.S. military operation to conquer alleged secret underground Nazi facilities in Antarctica and capture the German Vril flying discs, Thule mercury-powered spaceship prototypes.

According to Wikipedia, Highjump's objectives, according to the US Navy report on the operation, were:

Training personnel and testing equipment in frigid conditions;

Consolidating and extending the United States' sovereignty over the largest practicable area of the Antarctic continent (publicly denied as a goal even before the expedition ended);

Determining the feasibility of establishing, maintaining, and utilizing bases in the Antarctic and investigating possible base sites;

Developing techniques for establishing, maintaining, and utilizing air bases on ice, with particular attention to later applicability of such techniques to operations in interior Greenland, where

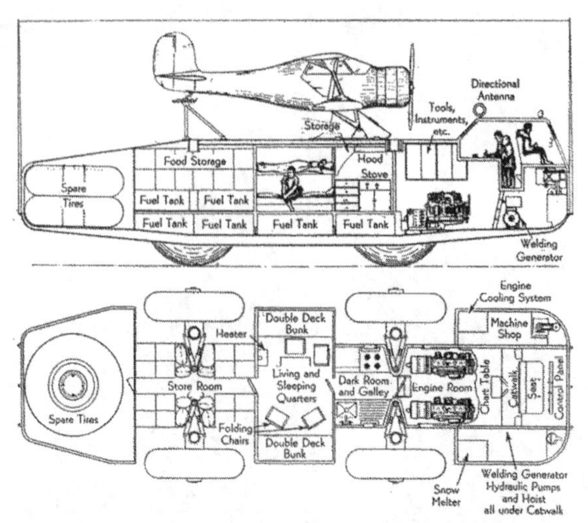

Diagrams of the Snow Cruiser and the attached seaplane.

conditions are comparable to those in the Antarctic;

Amplifying existing stores of knowledge of electromagnetic, geological, geographic, hydrographic, and meteorological propagation conditions in the area;

Supplementary objectives of the Nanook expedition (a smaller equivalent conducted off eastern Greenland)

Operation Highjump consisted of four groups of ships, an Eastern Group that departed from Norfolk, Virginia and a Western Group assembled in the Pacific. Both groups made it to Antarctica in December of 1946. A third group out of Norfolk was called the Central Group and it arrived at the Bay of Whales on the Antarctic coast on January 15, 1947, where they began construction of Little America IV. The fourth group was the Carrier Group (Task Group 68.4) with Byrd in charge from the aircraft carrier and flagship USS *Philippine Sea*. The USS *Mount Olympus* was the Communications and Flagship for the Central Group, commanded by Rear Admiral Richard Cruzen.

Only a month later the naval ships and personnel were withdrawn back to the United States in late February of 1947, and the expedition was terminated. The Navy said that this was due to the early approach of winter and worsening weather conditions, however, February was the height of the summer in Antarctica and the massive task force had originally planned to stay until August, another five months. Why did they abruptly terminate the mission? Did something happen in Antarctica that made the massive Navy group return to the United States to reassess and regroup?

There were some notable incidents, including some deaths. On December 30, 1946, the aircraft Martin PBM Mariner *George I* crashed during a blizzard which killed three of the nine crewmembers. The surviving six crewmembers were rescued 13 days later. In 2004 and 2007 attempts were made to recover the wreckage of this plane and presumably the frozen bodies of the three dead airmen. These attempts were unsuccessful and this wreckage has yet to be found.

On January 21, 1947, a seaman named Vance Woodall died during a "ship unloading accident" when he was crushed on the Ross Ice Shelf under a piece of roller equipment designed to "pave" the ice to build an airstrip.

Then in February of 1947 a fantastic discovery was made by US Navy Lt. Commander David E. Bunger when he flew over an area off the Shackleton Ice Shelf in Wilkes Land, and discovered an ice free area of lakes and land in a place later named the Bunger Oasis. Bunger Oasis, or Bunger Hills, named after the airman, is a coastal range on the Knox Coast in Wilkes Land consisting of a group of moderately low, rounded coastal hills, and is notably ice free throughout the year. It lies directly south of the Highjump Archipelago.

Flying over the ice and snow, Bunger spotted and landed on this 300 square mile patch of land that contains tricolor freshwater lakes that are totally ice free right in the middle of the blistering cold and freezing Antarctic! The lakes in the Bunger Oasis were the colors red, blue and green, from the vast amount of colored algae in them. So far, science has been unable to explain these lakes.

The Secret Land

The massive task force filmed a documentary about the expedition titled *The Secret Land* that was released in 1948. It was filmed entirely by military photographers from the Navy and Army and was narrated by actors Robert Taylor, Robert Montgomery, and Van Heflin. It begins with Chief of Naval Operations Fleet Admiral Chester W. Nimitz in a scene where he is discussing Operation

Highjump with admirals Byrd and Cruzen.

The film re-enacted scenes of critical events, such as shipboard damage control and Admiral Byrd throwing items out of an airplane to lighten it to avoid crashing into a mountain. In 1948 it won the Academy Award for Best Documentary Feature.

The Secret Land shows part of the fleet departing from Norfolk and traveling through the Panama Canal and then southward to their final destination. The trip through the ice pack is shown as fraught with danger and it forces the submarine USS *Sennet* to withdraw from the fleet or get crushed. The film says that the trip was a success and met all of its scientific goals. The film depicts the rescue of the six crew members from the crashed aircraft and the discovery of the ice free Bunger Oasis with the seaplane landing on one of lakes.

The aircraft carrier USS *Philippine Sea* had on its deck six giant RD4 supply planes and, with Admiral Byrd aboard, these flew over the frozen wastes of Antarctica and photographed them. This footage comprises much of the documentary including great shots of the Bunger Oasis. On one of Admiral Byrd's last flights he reportedly flew over the South Pole while speaking to the American

Transfering freight between the *Philippine Sea* and the *Northwind* during Highjump.

A lobby card for the 1948 movie *The Secret Land*.

public over live radio being relayed by the Navy to listeners all over the United States. However, during the broadcast, Byrd was cut off from his radio monologue for nearly three hours. The listening public thought that this radio silence may have been on purpose because of what Byrd was describing, and the Navy was deliberately killing the broadcast. Was Byrd describing flying saucers around his aircraft? Hollow earth believers thought that he was describing the famous "hole at the pole" that would take the aviator to the interior of the earth.

The official explanation, one recreated in the movie, was that Byrd was on his way to the pole and had to fly over high mountain ranges, and that Byrd had to jettison much of his equipment to avoid losing altitude and slamming into the side of a mountain. The Vinson Massif in Antarctica does have mountains up to 16,000 feet (4,892 meters).

Apparently, even though Operation Highjump was conducted over fifty years ago many of the photographs and documents concerning that expedition are still classified. We still are not sure what happened during Operation Highjump, and photographs circulate of flying saucers over the Little America base. It is rumored

A lobby card for the 1948 movie *The Secret Land*.

that Byrd's group even had an aerial battle with flying saucers over the icy wastes and such a battle may have been why one of the planes crashed.

Then, upon the return journey to Panama and the United States, Byrd discussed Operation Highjump in an interview with Lee van Atta of the International News Service that was held aboard the expedition's command ship the USS *Mount Olympus*. The interview appeared in the Wednesday, March 5, 1947, edition of the Chilean newspaper *El Mercurio* and read in part as follows (in Spanish):

> Admiral Richard E. Byrd warned today that the United States should adopt measures of protection against the possibility of an invasion of the country by hostile flying objects coming from the polar regions. The admiral explained that he was not trying to scare anyone, but a bitter reality that in case of a new war the continental United States would be attacked by flying objects which could fly from pole to pole at incredible speeds. ...I have to warn my compatriots that the time has ended when we were able to take refuge in our isolation and rely on the certainty that the distances, the oceans, and the poles were a guarantee of

A lobby card for the 1948 movie *The Secret Land*.

safety.

Another of the famous quotes from Admiral Byrd from his 1947 interview was, “I’d like to see that land beyond the Pole. That area beyond the Pole is the center of the great unknown.”

Shortly after making this statement, Byrd was hospitalized at the same Naval hospital in Bethesda, Maryland that James Forrestal was to die in two years later, and was not allowed to hold any more press conferences. He seemed to be considered a loose cannon on deck and not much was heard of Admiral Byrd for some years, and then in March 1955, he was placed in charge of Operation Deep Freeze.

Operation Highjump was followed up with Operation Windmill, an exploration and training mission to Antarctica in 1947–1948. The expedition was commanded by Commander Gerald L. Ketchum, and the fleet was called Task Force 39 with the flagship icebreaker USS *Burton Island.* It included the icebreaker USS *Edisto* which made a rendezvous with the *Burton Island* for the expedition. Missions during Operation Windmill included “supply activities, helicopter reconnaissance of ice flows, scientific surveys, underwater demolition surveys, and convoy exercises." It was also to provide ground-truthing to the aerial photography of Highjump.

Operation Deep Freeze

Over five years passed without any (known) US Navy incursions to Antarctica. If a German base existed there it was apparently completely left alone during these years. Then in 1954–55, the icebreaker USS *Atka* made a scouting expedition for future landing sites and bays for coming Navy operations. Byrd had suggested that the term "Deep Freeze" be used for all US Navy operations in Antarctica, therefore the upcoming expeditions would be Deep Freeze 1, Deep Freeze 2, and so on. The first Operation Deep Freeze was from 1955 to 1956. Admiral Byrd returned in 1955 for Operation Deep Freeze I and continued to make odd statements to the press.

In an *International News* dispatch on April 6, 1955 Admiral Byrd was quoted as saying that it was his mission to establish a satellite base at the South Pole:

> Byrd to construct Navy base on South Pole expedition. The Navy announcement said that five ships, fourteen planes, a mobile construction battalion with special Antarctic equipment and a total of thirteen hundred and ninety-three officers and men, will be involved in the expedition. …The expedition shall procure a satellite base at the South Pole.

Then, on November 28, 1955 Byrd made this statement before departing for Antarctica: "This is the most important expedition in the history of the world."

During Operation Deep Freeze the Antarctic bases at McMurdo Sound, the Bay of Whales and at the South Pole were established. Byrd, as usual, made a number of curious statements to the press. Byrd made a radio announcement on February 5, 1956, and said:

> On January 13 members of the United States expedition accomplished a flight of 2,700 miles from the base at McMurdo Sound, which is 400 miles west of the South Pole, and penetrated a land extent of 2,300 miles beyond

the Pole.

Then Byrd stated on March 13, 1956, "The present expedition has opened up a vast new land." Admiral Byrd continued to paint Antarctica in dreamy terms to Americans, stressing the importance and mystery and—bizarrely—danger that this land of the pole held. Then in 1957, just prior to his death, Byrd called Antarctica "… that enchanted continent in the sky, land of everlasting mystery." Byrd died less than a year after returning from Antarctica on March 11, 1957 at his home in Boston. He is buried in Arlington National Cemetery.

During Operation Deep Freeze I, on October 31, 1956, US Navy Rear Admiral George Dufek and others successfully landed an R4D Skytrain (a modified Douglas DC-3) aircraft at the South Pole. This was the first aircraft to officially land at the South Pole and the first time that Americans had set foot on the South Pole. The aircraft was named *Que Sera, Sera* after the popular song of the time and is now on display at the Naval Aviation Museum in Pensacola, Florida. One has to wonder, if the Germans did have a base in Antarctica for years, whether one of their aircraft—a flying saucer or otherwise—might have landed at the South Pole before Dufek. Indeed, having a base at the South Pole and commanding the airspace above the South Pole was one of the missions of Operation Deep Freeze.

This important flight marked the beginning of the establishment of the first permanent base at the South Pole. Today this base is known as the Amundsen–Scott South Pole Station. It was commissioned on January 1, 1957 and the building was called "Old Byrd" and lasted about four years before it began to collapse under the snow. Construction of a second underground station in a nearby location began in 1960, and it was used until 1972 when it was dismantled because of cracks. Other structures were built and the South Pole station still exists today and is bigger than ever. We will discuss the modern-day activities at this base in later chapters.

The statements by Byrd in his old age added credence to the belief in the hollow earth and mystical worlds inside the earth. The folks who believed in such things were thrilled to hear the famous Admiral Byrd talk about a land "beyond the poles."

Ray Palmer, the publisher of *Fate* magazine, was inspired by

Byrd and hollow earth theories to write his December 1959 article, "Admiral Byrd's Weird Flight into an Unknown Land!" Palmer interpreted Byrd's peculiar remarks as evidence that the earth was hollow, and that Byrd had actually flown into it. Furthermore, flying saucers were coming from this polar region. Readers remembered some of the odd scenes in the 1948 movie *The Secret Land*. Was it possible that there was some conspiracy going on here?

Had Ray Palmer and others stumbled onto a mystery that had been going on for over a decade that involved Antarctica, UFOs, Admiral Byrd and secret underground (or under-ice) military bases? Were the last battles of World War II actually fought in Antarctica? Did the Nazis actually complete their development of flying disk technology and create a battalion of Nazi flying saucers that survived until years after the official surrender of the Third Reich? It seems fantastic, but as more and more information has leaked out over the years, there seems to be some credible evidence that the Nazis had established several bases in Antarctica before and during WWII, and that they had been experimenting with unusual aircraft designs, including discoid craft, flying wings and tubular, wingless craft.

Chapter Four

Nazis From the Hollow Earth

Time is a one-way street, except in the *Twilight Zone*.
—Rod Serling

As a young tween in the 1960s I saw the advert that regularly occurred in 60s magazines for Dr. Raymond Bernard's book *The Hollow Earth*. It promised to answer all sorts of questions:

- Why does one find tropical seeds, plants and trees floating in the fresh water of icebergs?
- Why do millions of tropical birds and animals go farther north in the wintertime?
- If is not hollow and warm inside the Earth at the Poles, then why does colored pollen dust the Earth for thousands of miles?
- Why is it warmer at the Poles than 600 to 1,000 miles away from them?
- Why does the north wind in the Arctic get warmer as one sails above 70° latitude?

Who was Dr. Raymond Bernard? Was the book just a tongue-in-cheek hoax, or did Dr. Bernard really do some research?

According to Walter Kafton-Minkel[17] in his book *Subterranean Worlds*, Dr. Raymond Bernard's real name was Walter Siegmeister. He was born in 1901 to a family of non-practicing Russian Jews in New York City. His father was a doctor, and, according to Kafton-Minkel, Siegmeister was unusually concerned with sex and the differences between boys and girls at an early age. Female menstruation, in particular, fascinated him. He had one brother, who was later in charge of the music departments in some of Brooklyn's schools in the 40s and 50s.

Siegmeister obtained a bachelor's degree from New York's Columbia University in 1924, and his master's degree and doctorate from New York University in 1930 and 1932, respectively. Demonstrating his interest in the occult, his Ph.D. dissertation was entitled "Theory and Practice of Rudolf Steiner's Pedagogy."

His secretary at one point, Guy Harwood, said that when Siegmeister was a biochemistry student in Germany (apparently sometime during his doctoral studies) he took great interest in the substance *lecithin*. "It had been discovered that lecithin had therapeutic merits, good for glands, nerves and brain. He returned to New York and got a chemist by the name of Smith to start preparing lecithin in syrup-liquid form (prior to this lecithin was processed in hunks like yellow cheese) to be sold by the naturopaths, etc. He wrote articles about the benefits of lecithin, which were published in health magazines.

" ...The Food and Drug Administration issued, through the post office, a 'fraud' charge against him for selling a nutritional product that [the FDA] claimed to be worthless of any therapeutic merits whatsoever. Walter eventually won his case. Today Soya Bean Lecithin is sold throughout America in health food stores and is recommended for its therapeutic merits."

At the same time the FDA brought the charges of fraud against Dr. Siegmeister, the lecithin proponent, the post office stopped delivery of his mail. Harwood says that it was "due to this stopping of his mail to the name of Dr. Walter Siegmeister he changed his name to Dr. Raymond Bernard... the name 'Bernard' was connected with relatives of his named Bernard." Perhaps the Raymond portion was a nod to *Amazing Stories* editor Ray Palmer.

According to Kafton-Minkel, Siegmeister moved to Florida in 1933 where he started a newsletter entitled *Diet and Health* and attempted to set up a health food colony near Lake Istokpoga with a real estate developer named G.R. Clements. Clements was also a health writer but, according to Kafton-Minkel, was a real estate "shyster."

Clements and Siegmeister sold swampland to those wanting to

grow papayas, avocados, pineapples and other semi-tropical fruits. Unfortunately, the land flooded regularly and was unsuitable for crops. Siegmeister and Clements were threatened with legal action, and Siegmeister decided that it was best to leave the country.

Siegmeister went to Ecuador in 1941 to escape civilization. He followed a friend named John Wierlo, who had left for South America in 1940. In Ecuador, Siegmeister and Wierlo attempted to found a new promised land at an isolated spot in the eastern part of the country

It was while in Ecuador that Siegmeister appeared in the May 9, 1943 issue of *The American Weekly*. The sensational article was entitled "Hope to Breed a Super-Race in Ecuador's Secret Jungles," and was written by J.M. Sheppard.

In the article, Sheppard reported that Wierlo, "a blond giant of 24 years… 200 pounds of solid muscle, clad in native garb after his two year preparation in the jungle," was to wed a woman named Marian Windish, a "24-year-old girl hermit who lived two years in the Ecuadorian jungle without clothes, cooked foods, weapons or medicines."

Also interviewed for the article was the man who brought them together to create a super-race, Walter Siegmeister. Said Sheppard of Siegmeister, "At 40 years of age, Dr. Siegmeister has the skin of a child, a complexion a movie star would envy and the most unusual eyes I've ever seen—brown, extraordinarily large and of such depth and fire that they draw one's attention inexorably. Yet the manner of the man is one of meekness and solemnity."

The article asserted that an Edenic society, based on a raw food diet and natural living, was to be built on the eastern slopes of the Andes and a new race would be created. But there was apparently trouble in paradise.

Wierlo later wrote that he refused to consent to the "super-baby" creation scheme, and that Marian Windish was actually married to another man, but Sheppard had wanted to use Wierlo in his photos.

Wierlo also charged that Sheppard had "photographed bearded Siegmeister with long hair down, in a robe, walking on water (with supports just below the surface) and other 'miracles.'" In an earlier edition of *The American Weekly*, Wierlo said, Sheppard and his photographer had "faked some stories about meeting Tibetan Masters with prayer wheels on a nearby mountain." According to Kafton-Minkel, these were the "same stories picked up by Vincent Gaddis for his *Tales from Tibet* series in *Amazing Stories*…"

Wierlo wrote, "With such perversion of facts, both Sheppard and Siegmeister were banned from the use of U.S. Mails, and Sheppard was banned by the Ecuadorian Immigration Department."

Kafton-Minkel thinks that Siegmeister had a genuine desire to set up a super-race colony in South America. Since that did not seem to be working out, he continued on his quest to establish healthier eating patterns. Siegmeister journeyed to California and sold health food and two books he had written, now using the pen name Dr. Robert Raymond since he had been banned from using the U.S. mail under the name Walter Siegmeister. The two books were entitled *Are You Being Poisoned by the Food You Eat?* and *Super-Health thru Organic Super-Foods*. Robert Fieldcrest, the original publisher, was intensely interested in banned health methods, and mentions Wilhelm Reich and his persecution in the foreword to the original 1964 Fieldcrest edition of *The Hollow Earth*.

Siegmeister then began to use the name of Dr. Raymond Bernard and settled in Lorida, Florida. Guy Harwood, starting in 1945 as his secretary, conducted the doctor's business from Jacksonville. He visited with Dr. Bernard at his home in southern Florida many times and often observed him working on several manuscripts.

Says Harwood, "He was working on some of these books he later published when I worked for him. He would type all day and, sometimes half the night on these manuscripts. [He would] eat baked sweet potatoes and a lot of kelp. When he took traveling-trips he'd cook up a lot of popcorn… or roasted corn and ate kelp with it."

Kafton-Minkel claims to have seen only two photos of Siegmeister, one taken soon after

Raymond Bernard in 1943.

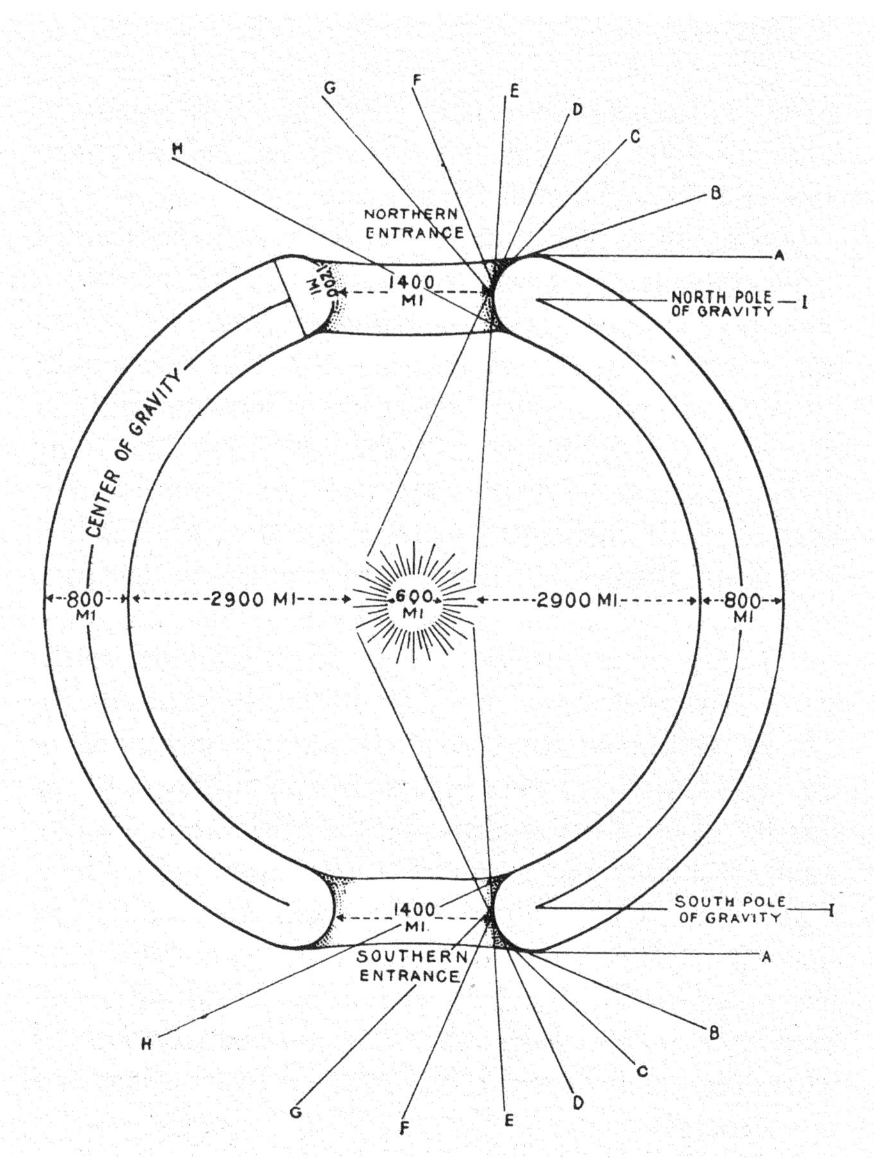

A diagram of the hollow earth showing the holes at the poles.

his graduation from NYU in 1933, and one which appeared in the *American Weekly* article in 1943. Both photos show a man with thick black hair and beard, and "penetrating eyes."

Siegmeister continued to travel for ten years, living at times in Hawaii, Guatemala and Puerto Rico, where he sold mail-order health books under the name Dr. Uriel Adriana, A.B., M.A., Ph.D. Kafton-Minkel reports that all this time he was searching for the right spot to create his super-race colony. Finally, his mother passed away and left him a sizable amount of money in 1955. Siegmeister moved to Brazil and purchased a fairly large bit of property near the town of Joinville on the island of Sao Francisco do Sul just off the coast of Santa Catarina state in southern Brazil. This is an area of Brazil that borders Argentina and Uruguay and has a large and prosperous German immigrant population.

Throughout his life Siegmeister had lived on the $200 a month that he got from his mother (not bad for the time, and a fortune in many foreign countries) plus the income generated from self-produced manuscripts that he photocopied or mimeographed and sold through the mail as books. He wrote the self-published book *Escape from Destruction* (1955) and sold it from Joinville, Brazil, again using the name "Dr. Raymond Bernard, A.B., M.A., Ph.D.," due to the mail ban.

In *Escape from Destruction,* Siegmeister, taking advantage of atomic bomb hysteria, wrote about the coming atomic war and how his colony in Brazil would be a safe place. He also mentions a Puerto Rican psychic named "Mayita" who predicted that a global nuclear war would begin between 1965 and 1970, and that by the year 2000 there would be no living thing left on earth. Extraterrestrials, however, would take the worthy to Mars where they would be safe.

Siegmeister/Bernard goes on to mention lost continents, Masters within the earth, and his belief that female human beings were superior to males. In his colony, sex was to be forbidden; women would ultimately regenerate through virgin birth, and eventually, they would all have to move to Mars with the space brothers. He also suggested that we might escape destruction by moving underground!

Flying Saucers from the Earth's Interior

About this time, Siegmeister, now firmly ensconced in his identity as Dr. Raymond Bernard, discovered a curious book published in Brazil in 1955 entitled (in English) *From the Subterranean World to the Sky* by O.C. Huguenin, a Brazilian writer who was apparently the director of the Theosophical Society in Brazil. Huguenin's book has never appeared in English, but Siegmeister translated large parts of the book after he obtained a copy in 1956 while browsing in a bookstore in Sao Paulo.

From the Subterranean World to the Sky relates Huguenin's thesis that flying saucers had been constructed over 12,000 years ago by Atlanteans, just before their continent sank into the ocean. Some Atlanteans used their craft to escape the destruction and migrate through the holes at the poles to the inner earth, where they reconstructed their advanced civilization.

Huguenin argued that the features commonly attributed to UFOs made little sense if they were extraterrestrial in nature, but were quite logical if they were from earth itself. The UFOnauts were monitoring radiation levels on the surface of the earth, he maintained, and had no intention of contacting governments on the outer earth.

Two other Brazilian Theosophists, an army colonel named Commander Paulo Strauss and a professor named Henrique de Souza, together described by Siegmeister as an "archaeologist and esotericist," were friends with Huguenin and provided him with much of the information in his obscure Portuguese book.

In 1956, Commander Strauss toured Brazil, lecturing on UFOs and the inhabitants' secret base inside the hollow earth that was called Agharta. He was presumably lecturing in an unofficial capacity, apart from the Brazilian military, though one has to wonder. He may have been on an official mission to disseminate false information on the very real UFO phenomena happening in South America, and all over the world, at the time.

Siegmeister indicated that Professor de Souza was in actual contact with Atlanteans of the inner earth, mentioning how residents of Sao Lourenzo frequently saw strange spacecraft landing near the Theosophical Society's headquarters located there; men "of great stature" would leave the ship, greet de Souza, and hurry inside for conferences of an esoteric nature.

According to Kafton-Minkel, Siegmeister paid a visit to the Theosophical headquarters in Sao Lourenzo to visit Professor de Souza. Siegmeister was publishing a newsletter from Brazil called the *Biosophical Bulletin*, and described the meeting in one of the issues:

> On the sofa in the back of the room sat a young girl, looking about 18 years of age. Much to my surprise she was introduced as [the professor's] wife, though he was over 70, and I was told that she is a subterranean woman, and really over 50 years of age, but retains her youth, since subterranean people live much longer than we do.
>
> The Professor began the conversation saying, "I just returned from a visit to the Subterranean World, where I am well known. I have frequently visited the city of Shamballah and once had the key of the door that leads to this city."

De Souza even told Siegmeister that he knew of a number of tunnel entrances in Brazil. One was in the Roncador Mountains of the Matto Grosso, the area where the famous British explorer Colonel Percy Fawcett disappeared in 1924.

The story of Colonel Fawcett and his ill-fated expedition to find Atlantean ruins in the remote jungles of Brazil was still very well known in the 1950s. The publication of Fawcett's son Brian's book *Expedition Fawcett* (published in the US as *Lost Trails, Lost Cities*) and the works of Harold Wilkins (*Secret Cities of Old South America* and *Mysteries of Ancient South America*) kept his fame alive.

De Souza told Siegmeister that Colonel Fawcett and his companions were still alive in the subterranean tunnels of the Roncador Mountains, in an Atlantean city. Said Siegmeister, they "are not permitted to leave lest they be forced to reveal its whereabouts." The entrance to the tunnels was guarded by fierce Chavantes Indians who would attack any intruder.

De Souza gave Siegmeister a password that would enable him to pass the Chavantes guardians and enter the tunnels. He told Siegmeister that the tunnel descended through several levels of subterranean cities and farms, but it finally ended in the great hollow at the center of the earth.

Siegmeister made trips to the Matto Grosso and Roncador Mountains, but could not find any subterranean entrances. He continued his self-publishing with a 1960 book entitled *Nuclear Age Survivors* in which he described himself as a searcher for the concealed entrances to Agharta.

Later in 1960, friends sent him copies of Ray Palmer's articles in *Flying Saucers* about the hollow earth and Byrd's flight "beyond the poles," as well as some of Theodore Fitch's pamphlets on the hollow earth and the free energy flying saucers. Fitch was an Iowa evangelist who believed in a hollow earth like the one described by Marshall B. Gardner in his 1913 book *Journey to Earth's Interior*—an inner world populated by "small brown men" who flew flying saucers powered by free energy and anti-gravity. Fitch maintained in his pamphlets that these inner world saints came out through the holes in the poles to occasionally observe humanity's progress on the outside. Siegmeister was impressed by this material and incorporated it all in his new photocopied books.

During this period Siegmeister wrote *Agharta, The Subterranean World* and shortly afterward, *Flying Saucers from the Earth's Interior*. *Agharta, The Subterranean World* was only 58 pages of self-typed manuscript, and *Flying Saucers from the Earth's Interior* was 89 pages in length. Both outlined Siegmeister's belief in a subterranean world, ancient continents and UFOs.

Siegmeister received a letter from a man named Ottmar Kaub, who said he was the secretary to Dr. George Marlo, the head of a foundation called World Research. Siegmeister published the letter in his *Biosophical*

Bulletin. The letter stated that Dr. Marlo was interested in living at Siegmeister's Santa Catarina monastery, was a vegetarian, and had "been taken inside the earth via the North Pole opening on many different trips on a flying saucer. He is in constant touch with the saucer people and in conference with them."

Ottmar Kaub, writing to Siegmeister continued, "because of his position and the orders he is under, he does not seek or desire any publicity whatever, and this letter is not for publication or showing to anyone outside of our group, because various governments and air forces do not want certain information given publicly."

Siegmeister and Dr. Marlo (through his secretary) exchanged letters and Dr. Marlo claimed that he had been in touch for the past several years with two beings named Sol-Mar and Zola.

Sol-Mar and Zola lived beneath South Africa in a city called Masars II. Marlo claimed that he had made over 60 trips with Sol-Mar and Zola in their flying saucers. Some of these trips were to the North Pole region where they entered the earth through a "curve" in the earth to the inner world, while other trips were to Masars II in the caverns beneath South Africa.

Dr. Marlo told Siegmeister that the inner world people called themselves "Terrans" and they lived in a lush world, much like our own, in which the sun never set. There were many cities, including Masars II, Eden, Delfi, Jehu, Nigi and Hectea. The inner world had an ideal climate, Marlo told Siegmeister, with people 12 to 14 feet tall, birds with 30-foot wingspans, and apples and oranges as large as a man's head. Obviously, they also had flying saucers.

The inner world people loved to cruise the outer world, Marlo told Siegmeister, and pick up "contactees" to confuse them as to the origins of UFOs. In a prelude to the "Bo" and "Peep" of the Heaven's Gate suicide group, Marlo claimed that Sol-Mar and Zola would pick up worthy candidates with their saucers and save them from the coming catastrophe.

Siegmeister writes about Marlo, Sol-Mar and Zola, all in glowing terms, in his books *Nuclear Age Survivors* and *Flying Saucers from the Earth's Interior*. Siegmeister used Marlo's material to seek more immigrants to his Brazilian commune, and suggested in advertising that a flying saucer shuttle would bring people between Brazil and the USA.

Over several years, Marlo, through the letters of Ottmar Kaub, continued to promise that he, Sol-Mar and Zola would arrive in Brazil in their flying saucer. However, usually agents of the US government stepped in to forbid the flights or he was suddenly called to Washington DC with his flying saucer.

Walter Kafton-Minkel says that Siegmeister finally realized that he had been hoaxed (perhaps willingly) by "Marlo" and "Ottmar Kaub." It may even be that Siegmeister fabricated this correspondence himself, attempting to get immigrants to his commune. Siegmeister was something of a trickster, but was fairly gullible as well. We will probably never know who "Marlo" and "Ottmar Kaub" were. Possibly even intelligence agents or the dreaded "Men in Black."

The Men in Black did appear to Albert Bender, the editor of a 50s UFO newsletter called *The Saucerian*, to warn him to stop publishing his newsletter and to tell him a bit about flying saucers. In Gray Barker's 1956 book *They Knew Too Much about Flying Saucers* it is related that Bender was visited by three men dressed in black suits who stated that the ultimate secret of UFOs was that they came from Antarctica. This was a popular book for its time, and both Ray Palmer and Walter Siegmeister would have been familiar with the claims made in the book. Where, indeed, many people wanted to know, were the saucers coming from? Were they coming from Germany? The Moon? Other planets? Military Bases? Antarctica? The hollow earth?

Siegmeister warned his readers in a late 50s issue of *Biosophical Bulletin*, "Don't come here expecting Dr. Raymond Bernard to bring you Atlanteans. He is hot on their trail, and must first meet them himself. After he does and he secures their permission, he will bring qualified refugees to them. The first step is to come to Santa Catarina, the New Holy Land. The next step is the tunnels."

By the late 1950s, Siegmeister's Biosophical community had stagnated. His island of Sao Francisco was home to only a few poor Brazilians and a few German immigrants from nearby Joinville. Says Kafton-Minkel:

> When his American correspondents wrote him of their searches for entrances to the inner world in the caves and

mountains of Arizona, New Mexico, and northern Mexico, Bernard became especially disturbed; he pleaded with them not to explore the dero-haunted caverns of North America. He felt the dero were the 'outcasts and degenerates' expelled from the Lemurian 'Motherland' thousands of years ago, and their presence beneath America explained the growth of juvenile delinquency in American cities. Desperate to convince his American readers to join him in Brazil, he went to work on his final book, *The Hollow Earth*.[17]

Nazis From the Hollow Earth

In compiling *The Hollow Earth,* Siegmeister used material from William Reed's book *The Phantom of the Poles* and Marshall B. Gardner's book *Journey to the Earth's Interior*. He also drew heavily from F. Amadeo Giannini's 1959 book *World's Beyond the Poles* and Palmer's Decembe, 1959 article in *Flying Saucers*, both of which concerned Byrd's expedition to the South Pole, but which Giannini and Palmer mistakenly said was to the North. Siegmeister later claimed that Byrd had made a "secret" trip to the North Pole in 1947, rather than admit the mistake. Conspiracy theories abound regarding Operation Highjump and its ultimate mission in Antarctica. Says *Wikipedia*, "Operation Highjump has become a topic among UFO conspiracy theorists, who claim it was a covert U.S. military operation to conquer alleged secret underground Nazi facilities in Antarctica and capture the German Vril flying discs, Thule mercury-powered spaceship prototypes."

Giannini, one of Siegmeister's sources for his hollow earth book, believed in the "continuity of space"—an obscure theory that held that the physical universe was "continuous" and that other planets could be reached from the polar areas of the earth. Both Giannini's book and Ray Palmer's 1959 *Flying Saucers* article quoted heavily from Admiral Byrd's statements (as reported in newspapers) that seemed to show that something very strange was going on at the South and North poles.

As we have seen, Byrd made an apparent reference to flying saucers in an interview with Lee van Atta. He was quoted on March 5, 1947 by the *El Mercurio* newspaper of Santiago, Chile under the headline "On Board the *Mount Olympus* on the High Seas":

> Adm. Byrd declared today that it was imperative for the United States to initiate immediate defense measures against hostile regions. Furthermore, Byrd stated that he "didn't want to frighten anyone unduly" but that it was "a bitter reality that in case of a new war the continental United States would be attacked by flying objects which could fly from pole to pole at incredible speeds. (*Wikipedia*)

The back cover of the February 1947 issue of Ray Palmer's magazine *Amazing Stories* promoting a story about "Space Ships in Antarctica."

In researching this statement, bizarre as it is, it seems that Byrd did in fact do this interview (the only one, apparently, after Operation Highjump) but what he meant by this could be interpreted in many different ways. Admiral Byrd's command ship in Operation Highjump was named *Mount Olympus*, so the interview apparently took place on its decks.

Giannini was particularly interested in Byrd's various statements to the press around the time of Operation Highjump. He quoted Byrd in an *International News* dispatch on April 6, 1955 as saying that it was his mission to establish a satellite base at the South Pole, the literal second invasion of Antarctica by the Americans:

> Byrd to construct Navy base on South Pole expedition. The Navy announcement said that five ships, fourteen planes, a mobile construction battalion with special Antarctic equipment and a total of thirteen hundred and ninety-three officers and men, will be involved in the expedition. …The expedition shall procure a satellite base at the South Pole.

Giannini also quotes a November 28, 1955 statement that Byrd made before departing for Antarctica: "This is the most important expedition in the history of the world." Then Giannini quotes a radio announcement made by Byrd on February 5, 1956, when he said, "On January 13 members of the United States expedition accomplished a flight of 2,700 miles from the base at McMurdo Sound, which is 400 miles west of the South Pole, and penetrated a land extent of 2,300 miles beyond the Pole."

Byrd had gone to Antarctica as part of the US Navy Operation Deep Freeze in 1955-56 as preparation for the important multinational expeditions to take place in Antarctica during the International Geophysical Year of 1957-58. During Operation Deep Freeze the Antarctic bases at McMurdo Sound, the Bay of Whales and at the South Pole were established. Byrd, as usual, made a number of curious statements to the press.

Giannini says that Byrd stated on March 13, 1956, "The present expedition has opened up a vast new land." Then in 1957, just prior to his death, Giannini says that Byrd called Antarctica "... that enchanted continent in the sky, land of everlasting mystery."

A Land Beyond the Poles

For Giannini, Byrd's unusual statements added credence to his belief that there was a land "beyond the poles," and helped prove his belief in the "continuity of the universe." Others, however, like Siegmeister and Ray Palmer, saw Byrd's strange statements as evidence that there were indeed "holes at the poles" and that flying saucers were coming out of these holes. Palmer had received a copy of Giannini's book in mid-1959 and was inspired by it to write his December 1959 article, "Admiral Byrd's Weird Flight into an Unknown Land!" Palmer, and now Siegmeister as "Dr. Raymond Bernard," were to interpret Byrd's odd remarks as evidence that the earth was hollow, and that Byrd had actually flown into it.

Had Ray Palmer and Dr. Raymond Bernard stumbled onto a mystery that had been going on for over a decade that involved Antarctica, UFOs, Admiral Byrd and secret underground (or under-ice) military bases?

Working with the fragmented pieces of the puzzle that they were given, Palmer and Siegmeister/Bernard formulated their now famous pop culture weirdness—the paradigm-busting belief that the earth is hollow, has holes at the poles, and that flying saucers are coming out from the center of the earth through these openings. Books and magazines on strange goings on inside the earth were hot sellers. Palmer had made a small fortune off the tales of Deros and Teros inside the caverns of Richard Shaver's strange rants in *Amazing Stories*, and the public loved books such as *Journey to the Center of the Earth* and *Tarzan at the Earth's Core*.

But did Palmer or Siegmeister suspect the bizarre truth behind their story: that the US Navy (who are also an Air Force) and the remnants of Hitler's Third Reich were still fighting it out in Antarctica? And, incredibly, these battles involved flying saucers and other craft deemed "UFOs." These flying saucers were also active in South America, where Siegmeister was now living, and the Americans, in connection with the British and Canadians

The December 1957 issue of Ray Palmer's magazine *Flying Saucers* with articles entitled "Admiral Byrd's Weird Flight into an Unknown Land" and "Polar Exploration Proves Saucers from This Earth!"

were building their own version of these craft in California, New Mexico and Nevada. Ray Palmer may have had some hint as to the man-made nature of some UFOs—and their connection to Antarctica—but it was safer and more profitable to promote UFOs along with extraterrestrials and the hollow earth. In this way, he would appear to be basically a "harmless, nut-job." And in those days, as long as you weren't a communist, you could largely keep your profile below the radar.

Walter Siegmeister/Dr. Raymond Bernard was another story. He basically was a communist, but had thoughtfully fled the US years before the communist witch hunts, and was so far out in left field that, except for the postal authorities, no one cared much about him. Even when *Flying Saucers from the Earth's Interior* was privately printed and sold by Siegmeister, the intelligence officers at the Air Force probably didn't pay much attention to

Flying saucers were all the rage in the 1950s as can be seen in this 1959 depiction of the soon-to-be world of personal flying disks.

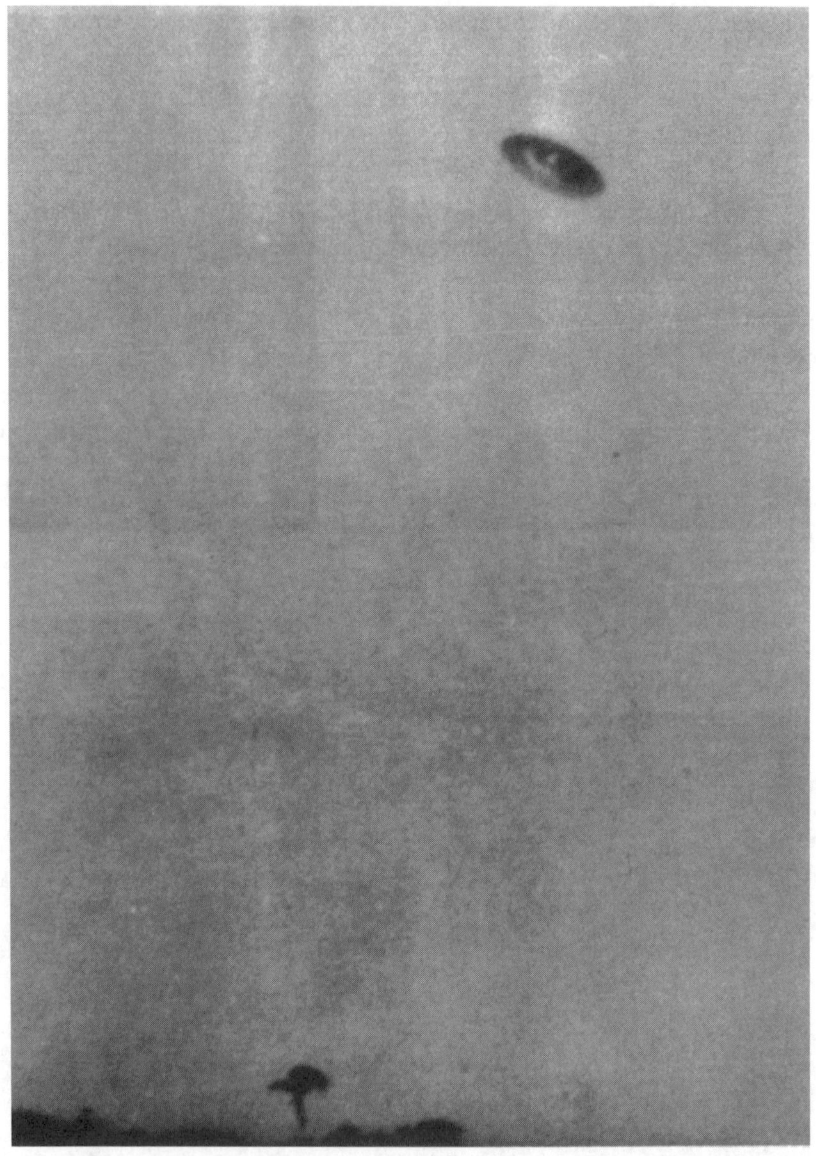

A photograph taken of a flying disk at Barra da Tijuca in Brazil, May 7, 1952.

him.

Siegmeister, now inspired by Ray Palmer's article in *Flying Saucers*, began the book that combined and refined his two latest passions—the hollow earth theory and flying saucers with advanced propulsion systems. But *The Hollow Earth* was unlike Siegmeister's earlier mimeographed "books," in that he was able to get a New York publisher named Fieldcrest (no longer extant), to publish his work in 1964. This was after he had circulated the book, much of it a rewrite of *Flying Saucers from the Earth's Interior*, as a mimeographed or photocopied private edition. *The Hollow Earth* was heavily advertised by Fieldcrest in 60s magazines. It sold quite well as a mail order book, introducing an entire new generation to the concept of the hollow earth. Fieldcrest sold out to University Books of New York who re-copyrighted the book in 1969.

Meanwhile, Siegmeister died in 1965 on his island in Brazil, reportedly of pneumonia. Kafton-Minkel reports that letters sent to him beginning in the summer of 1965 were sent back to the senders with "deceased" (in Portuguese, presumably) stamped on the envelope. Was Siegmeister/Bernard actually dead? We have no way of knowing. Like others, he may have outlived reports of his death.

The Hollow Earth goes beyond other books of its genre by suggesting that flying saucers were coming from inside the earth, rather than outer space. In fact, the first chapter of *The Hollow Earth* isn't about the hollow earth at all but rather "UFOs and Government Secrecy" as it is entitled. Siegmeister made the rational point that if there were holes at the poles—and flying saucers were coming from this interior earth as he claimed—then the US government must be covering up this incredible fact, along with the origin of the flying saucers. That the UFO enigma is a giant military cover-up is an argument that still continues to this day, more than 50 years later.

Flying saucers from the earth's interior were the great secret of our time, claimed Siegmeister, and these craft were being kept secret by the US government (actually a secret military-intelligence community within our government that controlled such black projects) and all the major world powers. Here we see the beginnings of a conspiracy that claimed that the US government and Russians were working together, as Albert Einstein had suggested in a famous 1945 letter to President Truman; the Cold War was largely a hoax in which the Russians and Americans were secretly working together to "divide and conquer" the rest of the world that hadn't been actively participating in WWII. The last chapter of *The Hollow Earth* is about free energy devices and anti-gravity, kept secret by "the secret government."

Did Siegmeister have special information on Nazis and Antarctica from his contacts in South America? He apparently had high-level communications with people in the Brazilian and Argentine military. Were his wild hollow earth stories just a cover for the activities of Nazis in South America and Antarctica? A cover story that we might call "the Nazis from the hollow earth."

Chapter Five

The Black Fleet

In this war there does not exist victors or vanquished,
only the dead and the living,
but the last battalion
will be a German battalion.
—attributed to Adolf Hitler

What seems clear from the odd reports coming out of Operation Highjump is that something strange was going on in Antarctica and it included submarines and possibly flying saucers. The researcher Henry Stevens concludes in his book *Dark Star*[46] that a small fleet of German submarines continued to move after the war between Europe, Greenland, a secret Nazi submarine base in the Canary Islands and ports in the southern Atlantic including Antarctica, Chile and Argentina.

Stevens also concludes that the bases used by the Germans in Antarctica were supplied from Europe initially and then later by German colonies in Argentina and Chile. He also concludes that these colonies were part of a secret manufacturing process that built flying saucers from parts that were specially imported from countries around the world, including the United States. Says Stevens:

> There is the idea that near the very end of the war in Europe, the German Navy refitted some damaged U-boats with non-conventional propulsion systems. This was unlike other developments by the German Navy, which involved extensive testing and prototypes and perhaps a whole line of test U-boats. These strange U-boats seemed to have been one-off and done on an individual and perhaps opportunistic

basis. After the war, these U-boats may have continued to function as sort of a phantom navy for the post-war SS.[46]

Stevens tells us that the first book in English to reveal information on the German saucer program and secret U-boats is by a California writer named Michael X. Barton. Barton's first book, *We Want You: Is Hitler Still Alive?,*[49] was published in 1960; a second book came out in 1968 called *The German Saucer Story*. It was with this second publication that German informants came to Barton and described a truly inside picture of events. Stevens gives us this quote from Barton's first book:

> Free-Energy, if released to this planet, could change things on earth as they have never been changed before. It could—undoubtedly—usher in a New Order of The Ages. Imagine it. A constant supply of free electrical energy taken right from the atmosphere by everyone. Yes, a universal power such as Free-Energy could release man from many of the burdens now present in our world money systems. But who has such a secret?
>
> The Nazis! If they have the original Viktor Schauberger "Electro-Magnetic" engine—or improved versions of it—they have Free-Energy. The UFO engine runs off power taken directly from the atmosphere—and that is quite inexhaustible.
>
> I have talked with a brilliant man from Europe. Holland, I believe, is his native land. It is, however, no effort for him to speak seven different languages fluently... including German. I asked him if he could give me any pertinent facts about the Nazis' use of Free-Energy, since I knew he had access to certain "inside information."
>
> "Yes," my friend replied, "it is known that the Nazis have a Free-Energy motor, and used it in 1958 to propel a U-boat between Europe and Buenos Aires. It is also believed by many of us that there is an underwater station—built by the Nazis—somewhere in the Atlantic Ocean between Germany and Argentina. A stopover place, no doubt, for Nazi subs and UFO's."[49]

Michael X. Barton with his wife at the 1962 Giant Rock UFO Convention.

This "underwater station" is apparently the secret U-boat base in the Canary Islands, to be discussed shortly. That there were a number of missing submarines after the official surrender of the Germans in World War II is a subject that generates a lot of interest within the military establishment. It is thought that up to 100 black U-boats might have been active immediately after the war. It is said that there were 35 U-boats participating in the still-secret final battle with portions of the British fleet somewhere near Iceland, on May 6, 1945 (approximately) in which the renegade U-boats sank all of the British ships, victims of the fury, during a daylong battle. This still-secret battle will be discussed shortly, but it is not the final battle of WWII; that battle was apparently fought in Antarctica.

The Mystery Subs

For years after WWII there were a number of strange incidents that included submarines of unknown nationality and origin. Some of these submarines were undoubtedly German U-boats that had not surrendered at the end of the war. These "mystery submarines" were part of what has been called the Black Fleet or the Ghost Fleet.

Says Henry Stevens:

> These are submarines of unknown nationality that were seen immediately following World War II, right up until this day where we call them USOs (Unidentified Submersible Objects). Of course, not all USOs are pirate German U-boats, the descriptions do not fit in every USO case. But for a few years after the end of the war this was certainly the case.
>
> Submarines of unknown nationality were especially prominent in reports from Norwegian fjords. This was the area of operation for the German U-boat fleet and waters they knew very well. What was happening was that a submarine of unknown nationality would be picked up by sonar or sighted off the coast of Norway, eventually being penned into a fjord there. NATO and of course the USA would jump on this, trying to force the unknown submarine to the surface with depth charges and other weapons. Sometimes this cat-and-mouse game would go on for weeks! It would always be reported in the European press and sometimes in the USA. Eventually and always, the submarine would slip away and the mystery would continue.

The U-218 leaving the port of Kiel, Germany in 1941.

> The problem was that this was the period of the Cold War. Obviously, NATO believed these were Soviet submarines and wanted to catch one for propaganda value. It was not absolutely clear that they were all Soviet in origin, especially in the beginning.

There is considerable evidence for German U-boat activity after the war, such as the story of a whaler called the *Juliana* (probably not the real name of the ship). A curious French newspaper article from *France Soir*, dated September 25, 1946 said (in French):

> One and one half years after the end of the fighting in Europe, an Icelandic whaleboat was stopped by a German U-boat. The whaleboat has been identified as the "Juliana" which was located in the waters near the island of Malvinas and the Antarctic zone, when it was stopped by a German U-Boat of magnificent size, bearing the German mourning flag (red with black borders).
>
> The commander of the U-Boat approached the whaleboat with a raft and gave orders to captain Hekla to surrender a part of the ship's provisions. The captain of the whaleboat had no other choice than to comply with the orders given by the German Navy officer. This officer spoke accurate English and paid for the provisions with U.S. dollars. In addition, he also paid a premium of $10.00 to each crewmember. The officer gave the captain of the whaler information pertaining to the location of whales. The ship later went to this described location and harpooned two whales. The rumors concerning U-Boats of the German Navy in the waters of "Fireland" and the Antarctic zone are now a fact (Telegram from AFP, dated Paris, September 25, 1946).[45, 46]

It should be noted that while this story has all the appearance of being true it seems likely that the names have been changed; there does not seem to be any Icelandic vessel that was ever named "Juliana" or an Icelandic captain named Hekla. Hekla is actually an active volcano in southern Iceland and most males in Iceland have a name that ends in "son." It would seem to be a "nom de guerre" for

A U-boat at sea.

the captain to call himself "Hekla." Still, this story has the ring of truth and is probably true.

Henry Stevens tells another curious story that happened at the very end of the war when the submarine U-534 refused to surrender as ordered to by Admiral Dönitz. Instead, she headed north from Denmark en route to Norway with two other U-boats that were unidentified. Says Stevens, quoting a website (now apparently defunct) about the U-534:

> May 5th, 1945: U-534 was sailing in the Kattegat, North-West of Helsingor. Although Admiral Dönitz had ordered all his U-boats to surrender as from 08:00 May 5th, U-534 refused to surrender. U-534—with two other U-boats in company—was heading north towards Norway, without flying the flag of surrender. Their departure was noted by Danish fishermen (?) and passed on to RAF Coastal Command that in turn sent out an air-patrol.
>
> A Liberator from 547 squadron attacked U-534. With all three U-boats firing at her she was shot down and crashed in to the sea. One survivor was rescued by a boat from the nearby lightship. By that time the two other U-boats dived, leaving U-534 alone on the surface.

At that point, says Stevens, a British surface ship attacked and sank the U-534 leaving three dead and 49 survivors. But the big story is that the other two U-boats got away! These two U-boats went through an attack unnecessarily, as they could have surrendered as Admiral Dönitz had commanded. The only possible conclusion is these U-boats did not want to surrender and were going to join the Black Fleet in Norway. We do not know the assigned numbers of these U-boats, and they could both be U-boats that were simply missing at the end of the war, or even U-boats that were listed as sunk—but were still active. Stevens said that by not surrendering and by instead diving to escape capture or destruction, they had become part of a post-war, Nazi, black U-boat fleet—the Ghost Fleet.

What follows is a list of all of the "missing" U-Boats as per the list on Wikipedia.

Name	Year Launched	Type	Captain
U-22	1936	IIB	Werner Winter
U-116	1941	XB	
U-122	1939	IXB	
U-184	1942	IXC/40	
U-193	1942	IXC/40	
U-196	1942	IXD2	E-F. Kentrat
U-209	1941	VIIC	
U-240	1943	VIIC	
U-337	1942	VIIC	
U-338	1942	VIIC	
U-355	1941	VIIC	
U-376	1941	VIIC	
U-381	1942	VIIC	
U-396	1943	VIIC	
U-398	1943	VIIC	
U-420	1942	VIIC	
U-455	1941	VIIC	
U-479	1943	VIIC	H-J. Förster
U-519	1942	IXC	
U-553	1940	VIIC	Karl Thurmann
U-578	1941	VIIC	
U-602	1941	VIIC	

U-647	1942	VIIC
U-648	1942	VIIC
U-666	1942	VIIC
U-669	1942	VIIC
U-683	1944	VIIC
U-702	1941	VIIC
U-703	1941	VIIC
U-740	1942	VIIC
U-745	1943	VIIC
U-851	1943	IXD2
U-855	1943	IXC/40
U-857	1943	IXC/40
U-865	1943	IXC/40
U-921	1943	VIIC
U-925	1943	VIIC
U-1020	1944	VIIC/41
U-1055	1944	VIIC
U-1191	1943	VIIC
U-1226	1943	IXC/40

This list amounts to 41 German submarines that are officially listed as missing and unaccounted for. Neither the Germans nor the Allies can account for these submarines, some of them later U-boats with advanced technology. And many on the list were probably retrofitted and altered near the end of the war for specific purposes, such as becoming passenger subs or cargo submarines carrying special materials such as large amounts of the liquid metal mercury. Some of these cargo submarines went as far as Japan with their special cargos.

We can add to this list of 41 officially missing German submarines some of the U-boats that are listed as sunk—and there are hundreds of these (too many to list here)—that were not sunk. Some submarines such as U-456 are officially listed as sunk but were apparently encountered after the war. There are also those whose numbers were given to another submarine so that there were in fact more than one submarine with the same number (as was apparently the case with U-530 as we shall shortly see).

Many of the German submarines had extremely long ranges

without ever having to refuel, and even those submarines that could not go for extreme distances without refueling would have been able to stop at secret submarine bases to refuel and resupply. These bases were located in Norway, Greenland, the Spanish Sahara, the Canary Islands, and even Antarctica. It seems that there might have been a secret submarine base in a fjord in the south of Chile as well.

The Secret Submarine Base in the Canary Islands

It is thought the Germans had a number of secret submarine bases outside of their regular bases in Germany, Poland, France and Norway. It is now known that there were secret bases in Greenland, the Canary Islands isle of Fuerteventura, and Antarctica. It is also thought that the Germans might have had secret bases in Argentina or Chile.

The secret base at Fuerteventura was key to secret submarine operations in the Atlantic, including voyages to Antarctica. It was known that some U-boats made cargo journeys of mercury to Japan during the war, and at least one known German U-boat was sunk off of Indonesia, its cargo was said to be tons of mercury.

The secret base on Fuerteventura is a fascinating story and gives us good insight into what a secret submarine base might be like—something straight out of a Hollywood movie. The island of Fuerteventura is a large, rocky island in the eastern part of the Canary Islands, and is the closest island to the African coast. This part of coast was the Spanish Sahara, an area that welcomed many ex-Nazis after the war. It was even rumored that Adolf Hitler was living in the Spanish Sahara (when it was still a colony of Spain).

According to Henry Stevens:

> There is no longer any doubt that a secret German base existed on Fuerteventura during the Second World War. One is reminded of the statement by Admiral Karl Dönitz about how the German Navy knew all the ocean's hiding places. Well, this is certainly one of them—a secret base in plain sight.
>
> This base has been mentioned by me in both my earlier books. Others have discussed this base, writing in the German language. For some reason general knowledge

of this base has not penetrated the consciousness of the English-speaking world, probably because of the language barrier. I say this because the real description of the base is in German that has never been translated into English.

Fuerteventura is a resort spot but an out-of-the-way one. From discussions I have heard there is a small tourist town there that includes a bar. Rumors of the secret German base have always been discussed and questions asked about it, mostly by tourists. Evidently, everyone or most everyone on the island denies these rumors in public but affirm them in private. But everyone, even you dear reader, have heard of this base in a roundabout way. This was the secret U-boat bunker featured in the George Lucas/Steven Spielberg film *Raiders of the Lost Ark*. In that film a U-boat sailed right inside the island, using a tunnel, into a huge cavern that was perhaps an ancient volcanic blister. There, the Nazi bad guys had erected submarine support equipment, turntables for U-boats, along with supporting manpower.

In 1971 *Stern* magazine published an article about the father of this base and the base itself in an article titled (and translated to English) "The Fantastic History of Don Gustavos, His Secret House and the U-Boat Base." According to the rumors and the facts on the ground, the father of this secret base was a German General, Gustav Winter. General Winter built a large, white villa on the high point of this island. The rumor was that a staircase descended down into the bowels of the island, connecting to this secret base.[46]

Stevens says that the German researchers Heiner Gehring and Karl Heinz-Zunneck did some research on General Winter, which revealed a few things about this secret U-boat base. Winter was born in 1893 in the Black Forest and was trained as an engineer. He died in 1971, a few months before the *Stern* article was published. General Winter performed some outstanding service for the Germans, and for that service he was granted land on the southern peninsula of Fuerteventura. During World War II, General Winter was the driving force behind a project to build a secret U-boat bunker on the island. He was also an agent for the Abwehr, one of the German spy services. The major feature of this bunker was a huge natural cavity that was connected to the sea by tunnels bored into the solid rock.

Ventilation shafts were dug upward towards what was now the large Villa Winter, which also included an airport. This facility functioned as a military base during the war and was heavily armed and guarded. After the war parts of the facility, with the exception of the airport, were destroyed with explosives. Today part of this large site belongs to a nature park but part of the property is still privately owned.

Stevens tells us that the best information comes not from *Stern*, but from another German magazine, *Nugget*. *Nugget* was a German-language magazine of treasure hunters. Below the title was written the words (in English) Gold, Minerals, Treasure Seeking,

Adventure. Stevens says that a two-part article was published in the July and August 1984 editions titled: "The U-Boat Bunker of Fuerteventura" (Parts One and Two). The article describes the adventures of two Spaniards and one Austrian who use a yacht and scuba gear to anchor off of the coast where the secret entrance to the submarine base was located and dive to the secret bunker.

The two divers (the Spaniards) slipped into the water on the ocean side of the yacht, so as not to be noticed by anyone on shore, and then swam underwater to the coastline that was essentially a volcanic cliff. A crack in the rock of this giant cliff led to a large natural cavern deeper inside the island. The Germans had evidently widened this tunnel-crack in the rock and reshaped the interior cavern, which was partly underwater and partially out of the water, and made three different submarine pens plus other rooms, carved out of the solid rock.

Two divers swam through the tunnel and then climbed up a metal ladder into the underground base. To their astonishment they saw that there were three submarine pens and two of them had submarines docked in the dark silence. They attempted to open the hatch of the first submarine and failed. They went to the second submarine and attempted to open the hatch. This time they succeeded.

Inside the submarine everything was neat and tidy. There was

A portion of the original Villa Winter on Fuerteventua in the Canary Islands.

no sign that there had been any trouble or that the crew had left in a haste. There were charts of the South American coast and other areas. Six points were marked on the coast of South America on one of the maps by a T with a circle around it. Between the various maps were newspaper clippings.

They left the piles of maps as they found them and exited the U-boat. They noted the numbers of both U-boats (but do not give them in the story) and later discovered that both of these U-boats were officially listed as sunk. They put their scuba gear back on and swam out through the underwater tunnel and back to the waiting yacht. They carefully climbed on board from the side that facied the ocean and not the island. The group then departed.

The article does not end there. The group continued to hang out in the Canary Islands, and the author says that they met several people who could tell them about the secret German U-boat base. They were told by one informant named "Charlie" that the base was fairly well known in the Canary Islands and that General Winter blew up the secret stairway that led to the U-boat pens in 1945.

The submarine bunker was still active until about 1950 they were told, though only accessible from the ocean. They were told that the base was known to the Americans, who allowed it to be

An aerial view of Villa Winter on Fuerteventura in the Canary Islands.

used, the informant said, to bring Project Paperclip Nazi scientists and officers secretly to the United States. They were told that there were originally three subs that came and went from the base but one of them was sunk near Florida—another missing sub, one that is probably on the list above. These submarines, and the base, were being operated by the SS during and after the war. The crews of the submarines were still living in the Canary Islands in many cases, though some had gone back to Germany.

Says Stevens:

> Here in the Canary Islands was an SS base that operated after World War Two. In fact, two U-boats were on-call and continued to be on-call for some time even after their last mission. We know this because the informant, "Charlie," includes the detail that the crew lived in the Canary Islands.
>
> This base functioned at least up until 1950 and with the knowledge and consent of the Americans. This would be the intelligence services of the USA, probably the CIA. We will get into this relationship later but we begin to see that this relationship between the surviving SS organizations, which became the "Nazi International" or The Third Power, was not always adversarial.
>
> Knowledge of this base and its continued existence after the war brings up claims of other facilities, laboratories, being operated by the SS after the war. Yes, there are those that say this is so and was done, not over the horizon in some never-never land, but right in Europe itself![46]

Stevens goes on to tell the story of a "Mrs. Maria W." who was aware of secret facilities in Western Germany after the war, including a secret SS research facility in the Jonas Valley. She went on to describe a small train track running to Bienstein (presumably also from the area of the Jonas Valley) that ran into a mountain. The Russians found the train track but did not ever follow the track fully into the mountain. This was the entrance to a German underground laboratory. For whatever reason, the Russians did not investigate the German facility there. "Mrs. W" says this was a "Fusionsanlage" an atomic fusion facility. Not fission but fusion, as in a hydrogen

A German train tunnel going inside a solid granite cliff.

bomb! And importantly, the woman says this facility operated until 1952! At that time it was supposedly shut down but remains ready to fully resume operation.

Stevens says that he sent copies of the Fuerteventura articles to researcher Mark D. Kneipp who wrote Stevens that he believed the two U-boats might have been U-1020 and U-1021. According to his work, these were both type VII-C-41 that "were completely rebuilt with the latest sound detection and weaponry just before the end of the war."

According to Wikipedia, the U-1021 is listed as sunk but the U-1020 is listed as a missing sub. So it may be possible that one of the subs was the U-1020, but the second (and third sub, allegedly sunk near Florida) may be other U-boats. However, some of the subs that are listed as "sunk" could have still been active after the war. Possible U-boats that are listed as sunk but could be in the "missing" category beside the U-1021 are the U-777, U-801, U-803, U-804, U-1017, U-1018, U-1051, U-1053 and many others. The last U-boat, U-4712 (built in 1945), was scuttled, as were most of the U-boat numbers before it. The last U-boat listed as sunk—U-3523—is known to have been sunk by depth charges from a British B-24

Liberator of 86 Squadron/G RAF about 10 nautical miles north of Skagen Horn, in the Skagerrak on May 6, 1945; all 58 crewmen were lost. The wreck was found by Sea War Museum Jutland in the 1990s at a depth of 123 meters. So this submarine is accounted for.

Still, it appears that there may have been as many as a hundred black U-boats in the Ghost Fleet that refused to surrender on the hour that Admiral Dönitz had proclaimed, though the figure is likely much lower—typically put at 35 for the submarines that assembled in Greenland at the end of the war, to be discussed shortly. The submarine base at Fuerteventura would have been well used as a safe intermediary base to go from such areas as Norway, Spain, and the remote Spanish Sahara (south of Morocco and controlled by Spain until it was ceded to Morocco in 1975) to areas further south such as Antarctica, South America, South Africa and beyond.

Stevens says that the existence of the base at Fuerteventura gives us a glimpse of what the other secret U-boat bases were like in Greenland and Antarctica:

> But the thing which should most concern us is the description of the base and its almost certainty of actual existence. If you accept the description of the base as a huge, hollow, natural volcanic bubble, connected to and accessible to the sea by tunnels, being full of machinery necessary to run a U-boat bunker, almost in plain sight, overflown and monitored on the surface by Allied aircraft and ships, then the proposition must be taken seriously that the Germans also had similar bases in areas where they would have been able to construct them unhindered. The fact is that the German base at Fuerteventura alone makes the case for the strong possibility if not the probability of similar bases in the Arctic and Antarctic.

Finally, Stevens tells us that the "outstanding service" performed for the German nation by General Gustav Winter, an engineer, for which he was given land on Fuerteventura and oversight of the base on the island was a curious invention. Before the war Winter worked on research that used strong magnetic fields to influence the compasses of ships and U-boats. This seems to have resulted in the Magnetofunk device that sends out strong magnetic signals that confuse the compasses and other parts of aircraft and ships. This Magnetofunk device was also used at the secret U-boat base in northeastern Greenland, another Arctic Base known as Point 103, said to be the Canadian Arctic, apparently around Baffin Island, and probably the secret Antarctic base known as Point 211. The function of the Magnetofunk device was to divert enemy aircraft that were seeking to find these secret bases in order to destroy them. The mysterious Foo Fighters over German skies at the end of the war apparently had a similar function: to interfere with the electrical systems and navigation of the Allied bombers.

The Strange Voyages of U-530 and U-977

Two known U-boats that operated with the Black Fleet in the months immediately after Germany's surrendered were U-530 and U-977. Both surrendered months after the war ended near Buenos Aires, Argentina. U-530 surrendered on July 17, 1945 while U-977 surrendered a month later on August 17, 1945. Both were Type VII U-boats. And clearly, they had both been doing something since the war in Europe ended on May 8, 1945. The officers and crews of both were accused of transporting Hitler and other Nazis to South America when they were interrogated, which was done first by the Americans and later by the British.

The German submarine U-530 was launched on July 28, 1942 and commissioned on 14 October 1942 with Kapitänleutnant Kurt Lange in command. The U-530 made six patrols. It departed the major U-boat base at Lorient in occupied France on May 22, 1944 for operations in the Trinidad area. On her outward voyage the sub made a rendezvous with the Japanese submarine I-52 and supplied this larger craft with a Naxos radar detector, an Enigma coding machine, a radar operator and a German navigator to help I-52 complete her journey. The I-52 is known as Japan's "Golden Submarine," because she was carrying a cargo of gold to Germany as payment for matériel and technology.

The I-52 was a Japanese Type C-3 submarine, designed and built by Mitsubishi Corporation between 1943 and 1944 as cargo carriers. They were quite long—longer than a football field— and carried a crew of up to 94 submariners. The Japanese constructed only three of these submarines during World War II (I-52, I-53 and I-55), although twenty were planned. These submarines were the largest submarines ever built at that time, before the enormous Sentoku subs were built, and were known as the most advanced Japanese submarines of their time.

It is interesting to note here that some of the photos of the German cigar-shaped craft show extremely long and thin craft—they appear to be of a similar design as the Japanese Type C-3 submarines—was one of these converted into an airship?

The two submarines rendezvoused on June 23, 1944 in mid-Atlantic, 980 miles west of the Cape Verde Islands and transferred the gold to U-530 and the secret plans and technology to I-52. The Allies had been informed of the rendezvous and directed the escort carrier *Bogue* and five destroyer escorts to the scene; *Bogue*'s aircraft managed to sink the I-52 with an acoustic torpedo. Although the sub was able to dive after being spotted, the newly developed Mark 24 "Fido" torpedo was able to track her and bring her down with over 100 on board and all the secret plans and technology just transferred—a total disaster for the Japanese.

Leaving the immediate scene, basically undetected, with the I-52 sinking, the U-530 returned to base, this time to the large U-boat base at Flensburg, Germany, after 133 days at sea. A short journey from Kiel to Horten Naval Base in southern Norway was her recorded next move, but it did not count as a patrol.

Now under the control of a different captain, Oberleutnant Otto Wermuth, who replaced Lange in January 1945, the U-530 was one of many U-boats that were based in Norway near the end of the war. The U-530 did not surrender on May 8, 1945 as did most of the German navy, but became one of the submarines of the Black Fleet. It is likely that U-530 went to the secret Greenland U-boat base where they met up with as many as 34 other U-boats—the Black Fleet assembled in Greenland was said to comprise 35 submarines.

What is known for sure is that the U-530 made a voyage after the war to the South Atlantic and Argentina, and the submarine probably stopped at the submarine base known as Point 211 in Antarctica. The submarine ultimately surrendered to the Argentine Navy on July 10, 1945 at the port of Mar del Plata. The captain, Oberleutnant Wermuth, did not say much. He did not explain why

The U-530 after it surrendered in Argentina in July 1945.

it had taken him more than two months to reach Mar del Plata, why the submarine had jettisoned its deck gun, what had happened to the ship's log, or why the crew carried no identification.

The unexpected arrival of the U-530 started a great number of rumors. Brazilian Admiral Jorge Dodsworth Martins said he believed that U-530 could have been responsible for sinking the armed Brazilian naval cruiser *Bahia*, which had sunk mysteriously a few weeks before. Brazilian Admiral Dudal Teixeira believed that U-530 had come from Japan, which is an interesting assertion.

It has been speculated that the Germans had a secret U-boat base in Indonesia or Melanesia, one that they would have shared with the Japanese, and which was still operational after the war. In many ways the early James Bond and Our Man Flint movies revolved around these secret bases, often on tropical islands, that were still operational after the war and were part of the Ratline, the Odessa, and even Project Paperclip (all operations to get Nazis out of Germany after the war).

It turned out after an inquiry that the *Bahia* had been sunk due to a gunnery accident by its own crew (ha, ha—with a deadly U-boat in the area a Brazilian warship sinks itself). But further rumors of mystery submarines and mystery passengers continued to surface in the South American press. An Argentine reporter claimed that he had seen a Buenos Aires provincial police report to the effect that a strange submarine had surfaced off the lower Argentine coast and had landed a high-ranking officer and a civilian who might have been Adolf Hitler and Eva Braun in disguise. Nazi gold was rumored to have been smuggled ashore to Argentina (and elsewhere in South America) and it should be remembered that when the U-530 met

The U-530 after it surrendered in Argentina in July 1945.

with the I-52 Japanese gold was transferred to the U-boat. Some of this gold may have ultimately been sent to South America.

Eventually, the Argentine Naval Ministry issued an official communiqué stating that U-530 did not sink the *Bahia*; that no Nazi leader or high-ranking military officers were aboard; and that U-530 had landed no one on the coast of Argentina before surrendering. The crew of the U-530 were interned by the Argentinian military as prisoners of war, as the United States had persuaded neutral Argentina to join the side of the Allies late in the war. The crew and the boat were then transferred to the United States military which brought the crew and submarine to the east coast of the US. The submarine was sunk as a target in a training exercise on November 28, 1947 by a torpedo from the American submarine USS *Toro*.

A couple of odd books have been written with input from a crewmember of the U-530, captain Wilhelm Bernhart, with his coauthor Howard Buechner. In these two books, written in the 1980s entitled *Adolf Hitler and the Secrets of the Holy Lance*[44] and *Hitler's Ashes*,[45] the authors claimed that the U-530 had stopped in Antarctica on its way to Argentina.

Buechner and Bernhart give us an unusual description of the German submarine U-530 that surrendered in Buenos Aires two months after the war in Europe had ended.

It is to be remembered that the U-530 was exceptionally fast. What kind of miraculous power plant propelled her on her swift course and what kind of wonderful fuel did she burn, which was efficient, safe, cheap and clean? Would the general use of this fuel have made the petroleum industry obsolete? And would it have unbalanced the economy of the world? Somewhere deep in the confidential archives of World War II, the answers to these questions await the coming of another era.[45]

Stevens tells us how this submarine almost certainly went to Antarctica before surrendering in Argentina:

> It is to be kept under consideration that one author of the above, Kapitan Wilhelm Bernhart, is an alias. In life he was a junior officer aboard this very U-boat. He ought to know what he is talking about. Kapitan Wilhelm Bernhart claims U-530 spent the time on a voyage to Antarctica and he claims to have made the voyage and the landfall in Neuschwabenland.

The U-977 made a similar voyage and probably went to Antarctica as well. The U-977 was launched in March 1943. The submarine was used in training and made no war patrols during its first two years of service. The submarine was sent on her first war patrol on May 2, 1945 sailing from Kristiansand, Norway. The ship was commanded by Oberleutnant Heinz Schäffer (1921–1979) who wrote a book, *U-Boat 977*,[43] about the voyage in German in 1952, with an English edition appearing later that year. Schäffer's orders were to enter the British port of Southampton and sink any ships there. Admiral Dönitz ordered all attack submarines to stand down on May 5. 1945, and at that time the U-977 was outbound north of Scotland from its home bunker in Norway.

It would seem that the U-977 was part of the Black Fleet that had earlier met in Greenland for one last battle with the British Fleet, as we will discuss shortly. Oberleutnant Schäffer in his book only said that he decided to sail to Argentina rather than surrender.

The U-977 after it surrendered in Argentina in August 1945.

During later interrogation, Schäffer said his main reason was a German propaganda broadcast by Goebbels, which claimed that the Allies' Morgenthau Plan would turn Germany into a "goat pasture" and that all German men would be "enslaved and sterilized." The Morgenthau Plan was a proposal, later scrapped, to remove all industry from Germany, leaving only agriculture, animal husbandry and service industries as the economy of Germany. It was an American proposal and it was ultimately resisted by the Russians and never passed. Certainly it was a radical proposal, one that would have alarmed a German captain and his crew.

Schäffer said that he offered the married crewmen the option of going ashore in Europe. Sixteen of his crew chose to do so and were landed from dinghies on Holsnøy Island near Bergen, Norway on May 10, 1945. Schäffer said U-977 then sailed to Argentina.[43]

After surrendering to the Argentine authorities, as had happened with U-530, the crew and the submarine were extradited to the US where they also responded to the charge of having torpedoed the cruiser *Bahia*. As we know, that sinking was later found to be due to a gunnery accident on the deck. The crew was then sent to the UK, where they faced accusations that they had landed Nazi leaders in Argentina before surrendering. Schäffer was held by the British as a prisoner of war and then released in 1947. The US Navy destroyed and sank the U-977, as they had the U-530, in naval firing exercises.

According to the US Navy report, the submarine stopped in

Captain Heinz Schäffer of U-977.

the Cape Verde Islands for a short break, and then completed the trip traveling on the surface using one engine, crossing the equator on July 23. Schäffer said that, after the short Cape Verde break, they completed the rest of the trip to Mar del Plata alternately on the surface and submerged. He did not say that they went to Antarctica.

The authors Buechner and Bernhart say that the U-977 made a rendezvous with another missing U-boat that was part of the Black Fleet:

> Finally, there is one more mystery about the U-977. It is said that, she too, made a rendezvous at sea with another submarine. In this case it was a soviet vessel carrying nuclear scientists and a very high-ranking Russian official. In exchange for German plans of a highly sophisticated neutron weapon, the Soviets agreed to block the implementation of the Morgenthau Plan that would have crippled post war Germany and reduced her forever to an agricultural state. This incident has not been verified.[45]

What made the U-530 and the U-977 surrender at all will always be a mystery. Kapitan Bernhart (not his real name) ultimately moved to the United States and supposedly lived somewhere in

The U-977 after it surrendered in Argentina in August 1945.

Missouri. It would seem that he was part of Project Paperclip—the secret OSS/CIA plan to bring former Nazi scientists and specialists to the United States after the war. Stevens mentioned earlier that the U-boat base at Fuerteventura was allegedly used in a joint operation between the CIA and the SS after the war, no doubt facilitated by SS General Reinhard Gellen. Also, we must acknowledge that all of these U-boat captains must have known about the Fuerteventura submarine base, as well as the secret submarine base in Antarctica, yet they never told this secret to anyone. The base in Antarctica did eventually make it into print, largely with Bernhart's account in the two books he co-authored with Dr. Howard Buechner.

More Missing U-Boats

A news dispatch from Reuters in 1998 claimed that another U-Boat, currently unidentified, was known to have been scuttled off the coast of Argentina after the war and two Argentinians were seeking to have the submarine refloated. Said the dispatch from Buenos Aires under the title "Refloat of Nazi sub off Argentina coast considered":

The U-530 and the U-977 in Rio de Janeiro's harbor being towed to the USA.

> Two residents of southern Argentina have asked permission to refloat a German World War II submarine sunk off the Patagonian coast, a German Embassy official said on Tuesday. German ambassador to Argentina Adolf Ritter Von Wagner "received the letter and it is in transit," an embassy spokesperson said. "We don't know what the decision will be. Antonio Rivera and Mirta Vicente pinpointed the location of the sunken German warship, which they claim to have seen in 1959, 1962 and 1966, newspaper *La Manana del Sur* reported. Area residents believe Germans disembarked from warships on the coast in 1945 near the end of World War II. Two other Nazi subs, a U-530 and a U-977, arrived in the port of Mar del Plata in July and August of that year. Argentina was a haven for Nazi war criminals such as Holocaust architect Adolf Eichmann, who was captured in Buenos Aires after the war, and Hitler's confidant Martin Bormann.[46]

This is a completely separate U-boat to those that had already surrendered such as U-530 or U-977. This U-boat was apparently scuttled at some period after the war ended and the crew went ashore in Argentine Patagonia and disappeared into the many German communities in Argentina, Chile and elsewhere in South America. Such German colonies are well known in South America, places like Colony Waldner on the banks of the river Parana not far from Buenos Aires where top Nazi Martin Bormann supposedly lived. Colonial Dignidad is another of these infamous German colonies, this one in Chile, though it was not founded until 1961. The Gestapo chief Gerneral Karl Heinrich Muller is supposed to have settled in Cordoba, Argentina, another Nazi stronghold.

Comments on a website about this story suggested that U-851, U-196 and U-1000 are possible candidates for being this unknown sub scuttled off of Argentina. Both U-851 and U-196 are listed as "missing," meaning that we have no idea what happened to these U-boats as they have completely disappeared from history. The submarine U-1000 is listed as scrapped in the East Prussian town of Pillau after it hit a British mine that had been dropped from the air by the RAF.

Stevens quotes from the German author O. Bergmann who sums up the Black Fleet (translating from German):

> At some time and some place, at secret U-boat bases outside the German motherland, U-boats of the German Navy must have been weighted-out, and, during the great withdrawal in April/May 1945, missing U-boats must have been equipped with new revolutionary technology and also must have been converted with electromagnetic propulsion. With this these [U-boats] may, [have been] arranged with the same possibilities and technologies as the German flying discs [called UFOs].[46]

The Refitting of Older U-Boats at the End of the War

It was common towards the end of the war for older U-boats and those that had been damaged to be refitted with new technology at the various U-boat pens in Europe. Says Henry Stevens in *Dark Star*:

> U-234 was converted from a minelayer to a cargo ship. It was originally a Type XB but three other Type IXD/1 U-boats, U-180, U-195 and U-219 were also converted to cargo carriers in 1944, with the express purpose of bringing high tech items to Japan. (1) U-234, it seems, could not wait for refitting as it was damaged by a bomb while still under construction in May 1943, so it was not launched until December 23, 1943. (2) In late August 1944, U-234 arrived in Kiel, German Werft for a refit. Twenty-four of the thirty-nine vertical shafts used to carry and lay mines were removed to make two large lateral holds. The other shafts were reworked to serve as cargo holds. A new radar unit with a small tower was installed, in addition to the latest snorkel model. (3) Upgrading old U-boats with a snorkel and modern electronics was an almost universal modification. Upon completing this refitting, U-234 left Kiel and formed up with three of the latest Type XXIII U-boats with the final destination of Norway using, at times, neutral Swedish waters. On a training voyage, U-234 collided with U-1301,

> puncturing the fuel tanks and causing a five-foot gash in her side. The dry dock facilities at the closest base, Kristiansand, was full. Captain Fehler was directed to Bergen, three days sailing time away. Rather than take this option, the Captain decided on a radical course of action. He flooded the forward tanks, raising the U-boat 20 degrees out of the water and having members of his own crew weld a piece of steel plate onto the skin of U-234. This took the better part of a week.
>
> In reality, this is like pipe welding. Pipe welding is done to a high standard as the pipe in question is subjected to pressure so that the weld itself must be done precisely and in a certain way. In the case of U-234, a long pipe, the precision of this weld or lack thereof carried with it the lives of all on board. The fact that Captain Fehler was willing to do this speaks of the competence of those crewmembers but also of the willingness and mental flexibility of the Captain and the ethos of the U-boat service. If he was willing to undergo a drastic field repair such as this and trust it going into action, then it is entirely feasible that other captains would be willing, also, to have their boats refitted in the field with untested, perhaps unauthorized machinery.[46]

Conventional German U-boats were powered by a combination of two types of engines. First, a diesel engine powered the U-boat while it was running on the surface. These diesel engines burned oxygen—usually, this meant atmospheric oxygen. While running on the surface, the diesel engine also ran a generator that charged the extensive battery array used to run the secondary electric motor. The electric motor does not need oxygen and so was used underwater or perhaps for escape when silent running was necessary. The downside to the electric motor was its range. Relying on batteries, it could only run as long as the batteries held electric charge. Once depleted, the U-boat had to surface or just lay there, dead on the bottom.

As the war progressed the U-boats were increasingly being attacked on the surface, even at night. The British and Americans were using radar to locate them on the surface and attack them from the air, seemingly out of the blue, in compete surprise. Surface running became dangerous for the Germans. But as things stood, it was necessary to run on the surface to charge the batteries in order to dive at all. In 1944 the Germans invented the Schnorkel, also called snorkel or "snort." This was a hollow tube that could be raised by the U-boat during shallow, underwater running, which connected atmospheric oxygen to the engine room for the diesel engines. Now, long distance voyages were possible and the batteries could be charged while the U-boat was technically under water. This snort could sometimes still be seen on radar but it was a big improvement. U-boat operation was again possible.

Various technologies competed for space aboard the new U-boat designs that came later in the war. The older Type VII and Type IX were slated for replacement with the Type XXI and Type XXII. A whole type, Type XVIIB, was created to test configurations for the Type XXVI. A great deal was riding on the development of new U-boats but the realization of a totally new, revolutionary, superior U-boat type as a total package, seemed beyond what was possible given the deteriorating war conditions. Says Stevens about the new technology and the retrofitting of it on older U-boats:

> All this experimentation documents the urgent need the Germans felt for an alternative to existing, conventional U-boat propulsion. But the official experimentation was bogged down in a bureaucracy and a convention of developmental prototypes. As with aircraft, what O. Bergmann was saying in the piece translated at the start of the U-boat discussion is that the Germans, at their individual U-boat pens, seemed to take it upon themselves to retrofit older U-boat models which had been damaged or damaged and sunken boats which had been refloated with new equipment, on their own, without official permission or orders, at least from the very top. This can certainly be said after the war ended since "official permission or orders" did not exist. The older models in question would have had to be Type VII as well as some Type IX U-boats.[46]

Some U-boats, including the U-530, were sometimes created with doubles. The U-530, was a Type VII c U-boat, Kapitan Bernhart tells us:

> At first glance the U-530 appeared to have been a highly unlikely choice for a long ocean voyage and the accomplishment of one of the most unusual missions of all times. She was old and leaky and had come to be known as an "unlucky" ship. In fact, she had recently been rammed by a freighter and at the time when her secret mission was to begin she was in dry-dock undergoing repairs [the author cites *National Police Gazette*, 1977].
>
> The mystery of the selection of the U-530 for special duty is easily solved. A new boat had been produced and equipped with improved engines, extra battery power, [a] schnorkel and other advanced features. For reasons known only to Admiral Dönitz she was assigned the duplicate number U-530 and hand picked by the Admiral for a highly secret mission. Above all else the new U-530 was fast. She is said to have been able to travel at 30 knots, even when submerged.[45]

The new U-530 left Kiel harbor on April 13, 1945, first stopping in Denmark to take on some supplies and then on to Kristiansand, Norway, arriving on April 20, 1945, to prepare for a long sea voyage. She departed on May 1, 1945 as part of a convoy of phantom ships, according to Buechner and Bernhart. And according to them the real U-530 would be found by the Allies in a dry dock in Kiel and appropriately entered into the records and duly accounted for. This meant the new U-530 was a black ship, with a duplicate number to one that was already captured. and was completely free to patrol the seas with no one looking for it. This fact would not be lost on the Allies after this U-boat surrendered.

Bernhart, while not mentioning the larger fleet, does mention that they rendezvoused with another Black Sub, the U-465 near Antarctica:

> Somewhere in the vicinity of the South Sandwich Island, the U-530 made contact with the U-465, another ghost ship which bore the number of a class VIIC vessel reported

> to have been sunk on May 5, 1943. She was now a type XIV known as an ocean going transporter or 'Milkcow.'[44]

The U-465 is supposed to have been sunk in the Bay of Biscay (the coastal area of western France and northern Spain) by an Australian aircraft in May of 1943. Forty-eight men went down with U-465 according to records and there were no survivors. With no survivors to tell the tale, the U-boat may not have actually been sunk as was believed; or the Germans had floated a new U-465 to confuse the Allies and create a Dark Sub that was already listed as sunk.

So, here we have another U-boat that is still in operation and not part of the "missing" list. From this meeting with the U-465, the U-530 is said to have traveled to Antarctica and deposited Nazi relics there. Bernhart claims to have been in the shore party. Although Bernhart does not mention it in his book, it would seem that they went to the German submarine base there.

And, Stevens quotes on the appropriation of a new type of U-boat by the US Navy in November of 1945 that is mentioned in a German publication:

> Actual proof has been uncovered that there could have been more here than only a design drawing. The US Naval

A U-boat crew preparing for their last voyage, 1945.

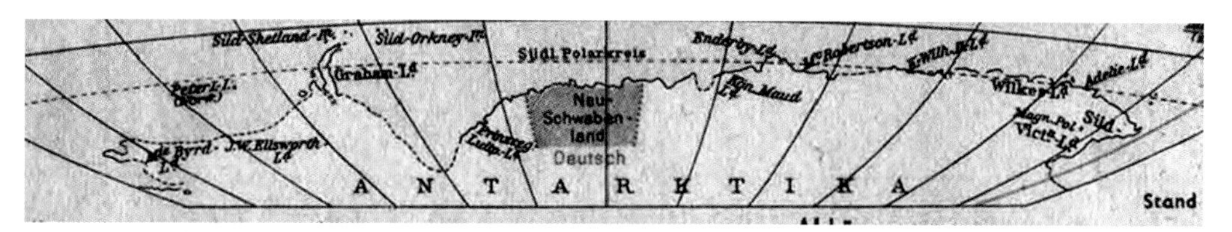

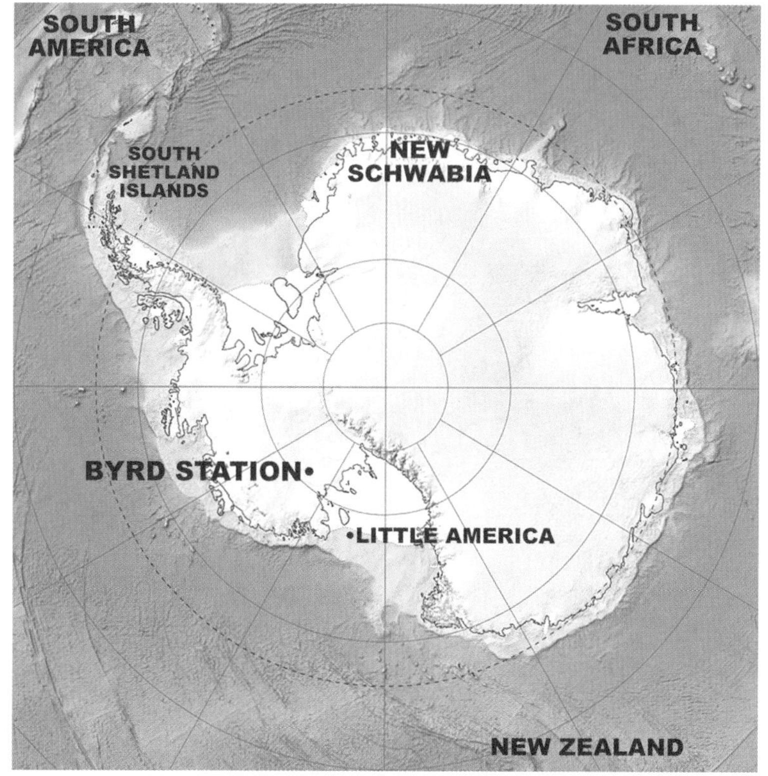

A map of Antarctica showing Neuschwabenland and other sites.

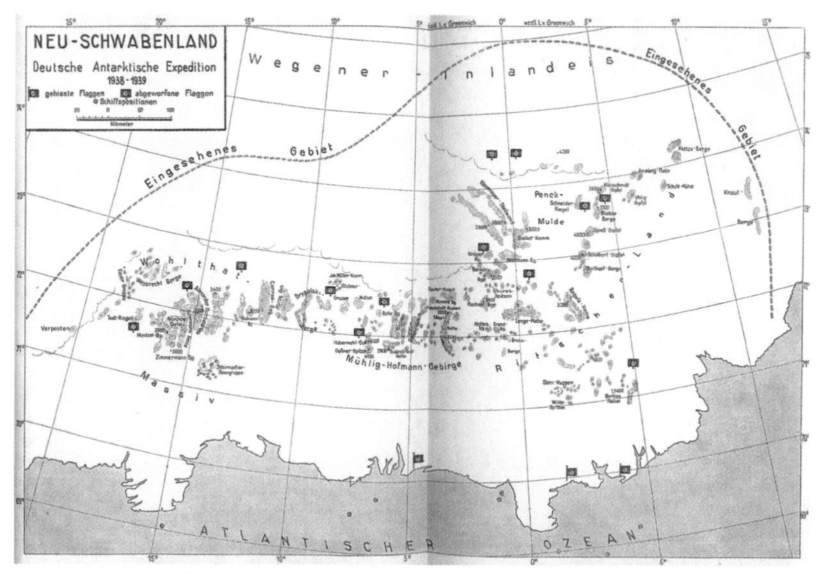

An early German map of Neuschwabenland.

A photo of the granite peaks of the Mühlig-Hoffman Mountains in Neuschwabenland. It is alleged that a secret base was built inside of this mountain range where saucer craft were stored.

The British Antarctic base at Port Lockroy at Wiencke Island in the Palmer Archipelago in front of the Antarctic Peninsula in 1962. The Palmer Archipelago and the South Shetland Islands contain numerous islands with granite mountains.

A map of Neuschwabenland and the Mühlig-Hoffman Mountains.

Left: The flag of Neuschwabenland as a German colony.

A map of Antarctica showing Neuschwabenland and other sites.

The former U-boat base at St. Nazaire, France showing the many submarine pens.

Above: A rare color photograph of U-977. Below: The Revell model of a Haunebu II.

Left: The special ring of a U-boat captain. *Below:* An SS Death Head (Totenkopf) ring and special ring box with the SS runes on it. As the sign of a secret society, it was to be worn on the third finger of the left hand.

An artist's concept of a Haunebu II in Neuschwabenland.

Above and below right: Two posters for the 1948 movie about Operation Highjump, *The Secret Land*.

Left: Admiral Richard E. Byrd on board the USS Wyandot during the first Operation Deep Freeze in Decenber of 1955. He died less than two years later.

Above left: A rare color photograph of a Haunebu on the ground. *Above right:* A rare color photograph of a Vril craft in flight. *Below left:* A rare color photograph of a Vril craft seen from below.

A photograph of a flying disk very similar to the Haunebu taken near Kanab, Utah on June 10, 1964.

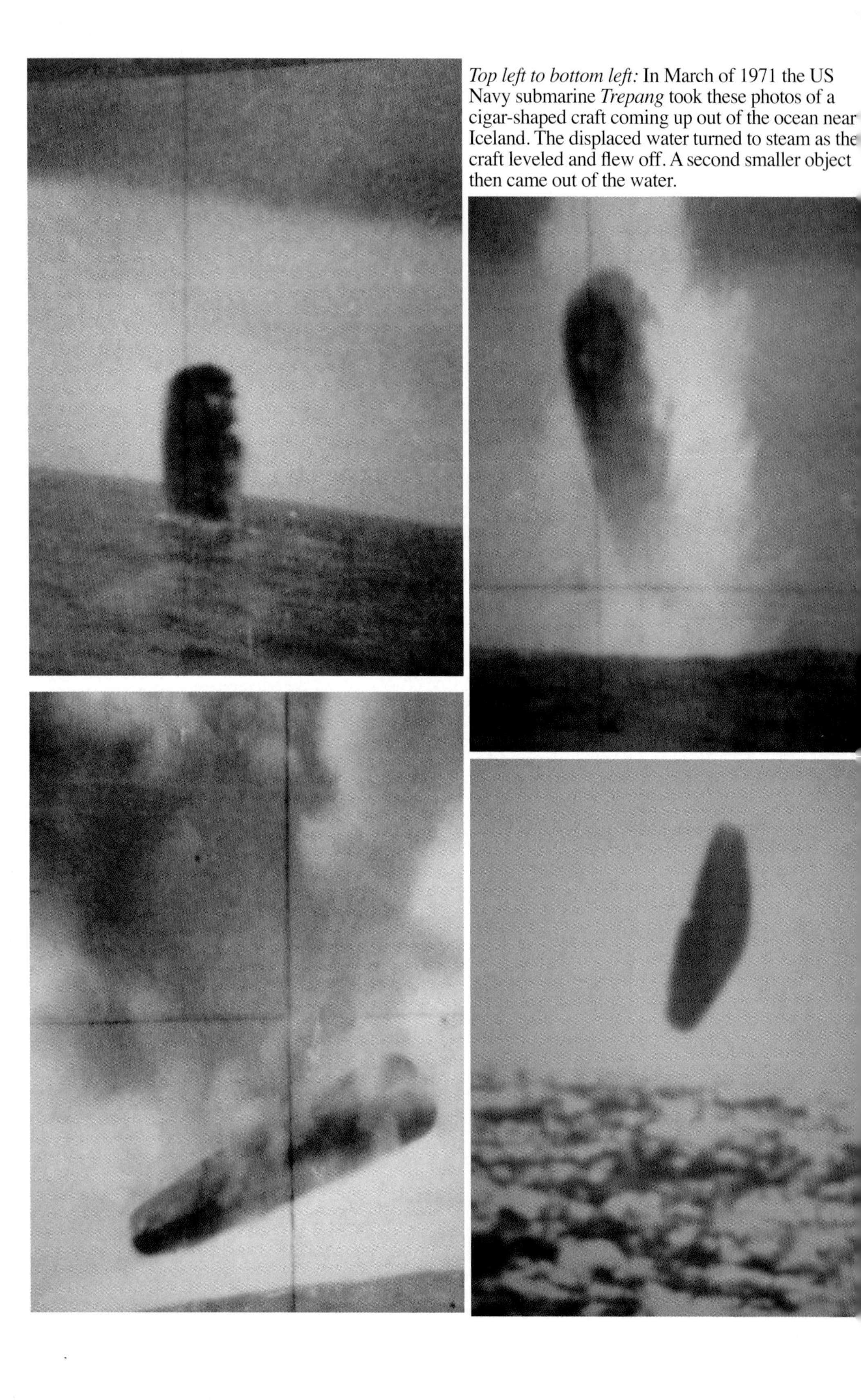

Top left to bottom left: In March of 1971 the US Navy submarine *Trepang* took these photos of a cigar-shaped craft coming up out of the ocean near Iceland. The displaced water turned to steam as the craft leveled and flew off. A second smaller object then came out of the water.

> Technical Mission had in November, 1945, on its own lifted a 195 ton German U-boat with a crane and freightered it to the USA. It is notable that no small German U-boat Type of this weight class is known. It is therefore very well possible that we are dealing with a research U-boat of the RPF with "(tear) drop form." A 195-ton water displacement fits very well with a length of 36 meters and a width of 5 m. In comparison, the Walter U-boat U-1406 displaced 415 tons and was more than double as heavy.[46]

This larger type of U-boat, the U-1406, was a research U-boat of a strange type. It was a rare Type XVIIb. The Type XVIIb, in total, consisted of three U-boats, U-1405, U-1406, and U-1407. These were built to test the feasibility of the upcoming Type XXVI U-boats. What they had in common was the Walter propulsion unit, which was a hydrogen peroxide power unit which had fallen out of favor, but was back again. The Type XVIIb was not an operational or attack U-boat, but just a test model. It could have been used as a cargo and passenger submarine, however.

Stevens tells us that the first advanced Walter units were not installed in a Type XXI, or a Type XXII, or a Type XVIIb, they were installed in the older Type VIIa units. Four of these were built, U-792, U-793, U-794, and U-795, all commissioned between November 1943 and April 1944. They were assigned to the 5th U-boat Flotilla but were retained for testing rather than seeing active, combat duty. All were scuttled on May 4 or May 5, 1945. Two of these submarines, U-792 and U-793, were raised by the British for more testing.

The Ghost Fleet and the Fuehrer Convoy

Colonel Howard Buechner and Kapitan Wilhem Bernhart describe the Ghost Fleet as the "Fuehrer Convoy" and state that up to 35 U-boats took part in a secret mobilization at the end of the war. These writers say that some of the submarines in the fleet made a stop at Kiel, in northern Germany near the eastern border with Denmark, on their way to Kristiansand in southern Norway, at which time many additional crew members were added. However, most of the torpedoes, ammunition and other armaments were stripped off these U-boats to make more room for cargo and personnel.

Once in Norway, these subs met up with others in the Black Fleet. On May 2, 1945 a large number of submarines left their Norwegian bases and headed toward the secret submarine base in Greenland that had the code name "Beaver Dam." They met up with submarines that were already on patrol in the North Atlantic and formed a flotilla of about 35 craft that were in Greenland on May 5, when Admiral Dönitz announced the surrender of the Third Reich. This flotilla headed south, toward the South Atlantic and Antarctica, and they were being tracked by a heavily-armed British flotilla.

On about May 6 or 7 the German Ghost Fleet and the British fleet met near southern Greenland and had a major battle that lasted for a day. In the end the British fleet was completely sunk and the German Ghost Fleet had triumphed in this battle. They had won the battle, but lost the war. The Ghost Fleet then continued south to the Canary Islands, Spanish Morocco, the Cape Verde Islands, to Antarctica, Argentina and beyond.

Stevens quotes from the German author, whose works have not appeared in English, W. Mattern, that the Black Fleet also called itself the Last Battalion and that Hitler once said: "In this war there does not exist victors or vanquished, only the dead and living, but the last battalion will be a German battalion."

Mattern calls this Hitler's end-of-the-war prophecy and—in the legend Mattern relates—Hitler escapes the end of the war in this U-boat convoy. According to the legend, Hitler left the bunker in Berlin and made his way to Norway. There, he boarded one of the U-boats, all with crews fanatically loyal to him, and the Fuehrer and a whole U-boat convoy sailed out of Kristiansand, Norway, on or about May 2, 1945. It may well be that Hitler was on this convoy, though he may also have died in Berlin, or, alternatively it has been reported that he went to Spain and the Spanish Sahara, Indonesia, Argentina and even Tibet.[33] More on the mysterious flights to Tibet in the next chapter.

After sailing out of Kristiansand the U-boat convoy met up with other submarines that had been on patrol in the North Atlantic and those based at the secret Greenland U-boat base known as Beaver Dam, and another one called Point 103 in the nearby Canadian Arctic. According to the German sources, the Black Fleet met up with a British Navy flotilla much as described above. On approximately May 6, give or take a day, an all-out naval battle took place somewhere north of Iceland near the east coast of Greenland. The British lost this battle—a battle never acknowledged by the British government—and the Black Fleet continued south to Antarctica with its valuable cargo.

The legend says that a large base was created from the earlier U-boat base, presumably the one at Point 211, with a population of younger men and women from the subs who had been specially chosen for this final mission. The submarines then dispersed to South America—we know the U-530 and the U-977 surrendered in Argentina—as well as to the secret bases in the Canary Islands, Spanish Sahara, Europe, Indonesia and even Japan. We will probably never know what fascinating missions these U-boats went on when they left, one by one, from the base or bases in Antarctica.

Stevens says that he came across another German author, Wilhelm Landig, and his book *Wolfzeit um Thule* and realized that this was the source for many of the stories told about the Black Fleet. Wilhelm Landig, a former member of the SS, carried forward the Aryan mythology of Thule and popularized the concept of the Black Sun, the mystical symbol that replaced the Swastika. He published *Wolfzeit um Thule* in 1980. This was Landig's second novel "full of realities." Landig actually opens the book with a description of this battle, the title of the chapter being "Die Verheimlichte Schlacht" (The Secret Battle). Landig tells us the story mostly through the interactive conversation of two co-captains, Captain Krall and Captain Hellfeld, aboard a supply U-boat in the U-boat flotilla. Each U-boat had two co-captains for some reason. The storytellers' U-boat is not given a proper "U" designation, only being called "Boot 5XX."

Landig tells us that the fleet of U-boats put out to sea from Kristiansand, Norway, on May 2, 1945, just as others have claimed, including Kapitan Bernhart. Landig calls it a "Geisterflottille," a ghost fleet. Among the U-boats assembled, Landig states that there was a Type XXI present. He also describes a totally different and totally new U-boat, not a Type XXI, Type XXIII or Type XXVI. Landig's book did not give any figures for the size of the ghost fleet.

The crews of these U-boats were not composed of standard

operating crews. Landig says that most were younger than 25 years old. Also, technicians and scientists were aboard. Some of these people, Landig says, were already listed as missing so they would not be sought after or expected to return. We know that the submarine U-530 was already listed as accounted for at the docks in Germany—the old U-530.

All of the crew of this fleet—men and women—left with the common understanding that the war in Europe was lost. They were not trying to change the course of the war and knew that Germany would be overrun by the Allies. They also understood that they were on a special mission, but did not know where they were going, which was Antarctica.

Landig mentions what we already know, in that most of these U-boats had been upgraded to some extent and some possessed the very latest in technology. For instance, all had the schnorkel device that would allow fresh air to enter the U-boat while submerged at a shallow depth. This made it possible to run on diesel power to recharge the submarine's batteries, necessary to operate the electric engines for the quiet underwater running, keeping them undetected.

Other devices mentioned by Landig are a round, dipole antenna used for warning the ship of an approaching enemy. Another device is called an "Ortungsgeraet" which pinpointed enemy ships within a range of eight nautical miles. Additionally, Landig lists the Balkon-Horchgeraet as being used. This device is said by Landig to have a range of 40 nautical miles. It picked up the sounds of propeller screws. The name "Balkon" (balcony) refers to the balcony-like housing built on the U-boat to house the sensing head of the device.

When the fleet was north of Iceland, about May 5, Allied radio chatter was encountered and the Germans knew they were in the area in some strength. But they also knew that the Allies were not expecting such a sizable German force in the North Atlantic at this late stage in the war. The Germans were confident of their superiority, based on the new technology and the great speed of their U-boats. They were not as heavily armed as they could be, but the U-boats still had plenty of torpedoes of various kinds to sink a large fleet. Apparently they also had an electromagnetic jamming device derived from Nikola Tesla's work.

Landig then says that a "flying object" from a German base

outside of the Reich (from the secret base located in northeast Greenland, "Beaver Dam," obviously) observed a group of the hostile ships in the distance and the information was relayed by long-wave radio to the U-boats. Stevens says that this is significant. He says that the Germans perfected long-wave communication during the war and that this was a type of Tesla technology in which long radio waves were sent by dipole transmitters through the earth. In this case "earth" also means the ocean so the Germans could pick up communications while submerged almost anywhere on the earth.

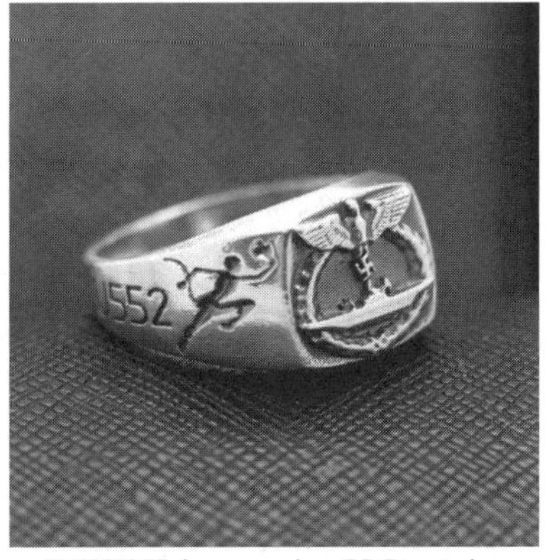

WWII Kriegsmarine U-Boat ring.

Various regions had layers of water which best facilitated reception of these messages and the depths for best reception where mapped out by the German Navy, worldwide. The transmitter required was very large and specialized, and needed to be on land. Transmissions could be received but not sent by the U-boats. It is thought that there was only one transmitter in existence and this was in Magdeberg, near Berlin. The facility was code named "Goliath." It seems that a second transmitter was located at the secret base in Greenland. Was there a third transmitter located at the Antarctic base of Point 211?

Stevens says the Goliath transmitter at Magdeberg had already been destroyed by the advancing Soviet army. Landig says the transmission did in fact come from the secret Greenland base and that the transmission was picked up at a depth of 20 meters. Landig also says that the "flying object" which made the original observation of the enemy ships was a German flying disc that he calls a "V-7" and says it was based at the Greenland base. Flying saucers over Greenland in 1945!

Landig gives us further interesting information when he tells us that the means of communication used by the V-7 was the "Kurier" (courier) device. The Kurier device relayed messages in a very unique way. Today, says Stevens, we call this method "burst transmission." A great amount of data was dialed into the Kurier device and transmitted in a burst of radio transmission so that the enemy could not get a fix on the sender. This was extremely successful for the Germans and Stevens says that after the war the first question asked of U-boat crews by the British concerned the Kurier device.

Landig says the battle with British flotilla started with the location of the fleet being detected with the V-7 while the U-boat fleet remained undetected. The U-boats then attacked the British ships with torpedoes in a devastating surprise attack by far more submarines than the British ever expected. Landig says that the weapon of choice for this attack was a particular kind of torpedo called the "Type TXI." He states that it operated at a depth of 50

A German U-boat concealed inside a protective slip, preparing for a long voyage.

meters, that it was steered by a sensor which homed in on the sound of propellers of the target ship, and that the ship could fire up to six such torpedoes at a time.

Stevens says this torpedo is a variation of the torpedo which was known as the king among German torpedoes, the “Zaunkoenig” or “T-5.” This torpedo was equipped with a magneto-striation transceiver that received, evaluated and then conveyed the direction from which the sound waves, issuing from the ship’s screw, had been received to the rudder through a gyro-device. The torpedo first ran in an arc and then turned in the direction of the sensed target. A safety device had to be built in to ensure that the torpedo kept sufficient distance from the U-boat and did not hit it.

Landig says that the provision and supply U-boat 5XX had no weapons aboard and could only sit tight and wait out the battle. Eventually the captains raised the periscope and had a look around. All the British ships were either sunk or sinking. No British ships survived. After the battle, contact with the V-7s in the air continued as was contact with the secret base in Greenland with its special transmitting station.

Then in Landig’s book the two captains on the unnamed U-boat discuss how the Americans will know there is a German presence in the Arctic and they will certainly and actively seek the German base(s). Comments Stevens:

> This brings up the postwar episode of Americans searching the Arctic for flying saucers. …retired Col. Wendelle Stevens was among those involved in this search in those days. He has stated that B-29s were fitted with photographic gear and patrolled large tracts of the Arctic. In some cases, according to his testimony, contact was made with flying saucers, both in the air and on the ice and they were successfully photographed. The film was taken off the B-29s and sent to Washington, D.C. undeveloped. Col. Stevens never was given any sort of explanation as to the photo-interpretive results.
>
> Returning to Landig and the captain’s talk of Arctic bases, it was stated that all Arctic bases would be moved and relocated to the Antarctic where the Germans would

> have more room and security. Besides this, the probability of Hitler's death was discussed, as was the probability of the Arctic Blue Island legend. No firm conclusions were reached on either. This concludes Landig's discussion of the last sea battle of the war.
>
> We should note that at no time does Landig use the words "Last Battalion." There is no mention of Adolf Hitler aboard the ghost fleet. Landig does mention a German flying disc but the only weapon mentioned by Landig was the sound-seeking torpedo.[46]

"Blue Island" is a reference to Point 103, which was seen in Canadian air patrols, to be discussed shortly. Stevens then moves on to the German writer W. Mattern, mentioned earlier, who wrote sometime in the mid-1980s, a few years after Landig's *Wolfzeit um Thule*. In his German-language book *UFO's Unbekanntes Flugobjekt? Letzte Geheimwaffe Des Dritten Reiches?* ("UFO's: Unidentified Flying Object? Last Secret Weapon of the Third Reich?") Mattern uses Landig's description of the sea battle and greatly expands on it. Says Stevens:

> This expansion is not in terms of proof or details but rather in terms of meaning. To Mattern, the sea battle between the ghost U-boat fleet somewhere near Iceland represents only a first adventure for what he calls "the last German battalion." It is his understanding that the German last battalion escorted Hitler—who did not die in the bunker—all the way to Antarctica to set him up in his new seat of power, Neuschwabenland. As supporting documentation, Mattern only cites the magazine *Der Weg* and *El Mercurio News,* but gives no date or other identifying data so that it can be verified. Landig's ghost fleet has now become the "Fuehrer Convoy."
>
> There is another important difference. While Landig uses German flying discs for reconnaissance, Mattern makes them the very instrument of Nazi power. They become the vehicle by which the Nazis will perhaps achieve redemption and reassert themselves in the world's mainstream at some

> future time. In fact, this is the point of the book. There is still no mention of other exotic weaponry being used by the U-boat fleet itself.
>
> ...My translation of Mattern's words are as follows: "It was perhaps Dönitz and not Goering who guaranteed the U-boat flotilla to the Antarctic and who also took over the security of the withdrawal of the Fuehrer from the bunker of the Reich's Chancellery, whereof Goering had no knowledge."

Finally, looking at the last of his German language sources, Stevens mentions another source, Norbert Juergen-Ratthofer, who adds more dimension to the still-secret battle above Iceland, which is an electric cannon:

> [There was] a citation from a magazine article Juergen-Ratthofer wrote. In the last sentence of the otherwise standard rendition of the Landig story, Juergen-Ratthofer describes a new type of torpedo but also a totally different weapon that he calls the "Electrokanone" (electric cannon).

The Secret Nazi Arctic U-Boat Bases

Before we can move to the subject of Nazi activities in Antarctica, South America and elsewhere, we must look at the secret Nazi bases, one in northeast Greenland ("Beaver Dam") and one in the Canadian Arctic somewhere on or near Baffin Island, called Point 103, and seen as a "Blue Island" by Canadian air patrols. Both of these bases were in contact with an air and submarine base at Banak in northern Norway. At Banak, long-range aircraft were based that were operated by the SS. Some of these planes allegedly had the insignia of the SS cult of the Black Sun on them.

Stevens mentions an article in the German *Mensch und Schicksal* magazine written by Claude Schweikhart in 1952, talking about a Nazi-hunting group called the Emerich-Team, the first investigative body ever to look into the esoteric Reich and the "Third Power." In the article Schweikhart makes mention of a "Laboratory of Death" in Greenland, probably at the base called Beaver Dam, and in another statement concerning flying U-boats the magazine said:

> Interestingly enough the Emerich-Team recorded in such documents, obviously working together with the Third Power-hunting Yugoslavian radio news service, the existence of a "Laboratory of Death" somewhere in Greenland which was designated a city of technical satanism and in 1951 is said to have tested a flying undersea boat.[46]

As the war drew to a close and Germany was losing, what was left of the German navy—largely submarines—essentially withdrew to occupied Norway. Norway contained a number of U-boat bases and with its many fjords was an ideal place to hide submarines and even conventional boats. Says Stevens:

> The central Norwegian coast is fjord country. These fjords harbored the German Navy during the war. At first it was the refuge of capital ships such as the Bismarck and Scharnhorst but later these facilities took on an even greater importance as U-boat bunkers in France were bombed and eventually overrun. The U-boat facilities in Norway became the home for the U-boat fleet and the U-boat fleet became the real German Navy.
>
> It was from this area that U-boats were sent out in the final days of the 3rd Reich on all manner of desperate missions. Until the very end, U-boats apparently awaited Nazi bigwigs who never showed up for their journeys of escape. Later, we will recount the mission of U-234 as well as the U-boats of the legendary "Last Battalion" whose point of departure was exactly these waters.
>
> But for now we must return to the springboard for the Nazi post-war world, the airbases of Northern Norway from which the Germans explored and watched over their holding in the Arctic. From here special long distance aircraft, including prototype aircraft salvaged from the jaws of Germany's collapse, were operational. Also, something new was in the air, a new identity was forming. Not only Wilhelm Landig tells us this in 1971, German engineer Claude Schweikhart (sometimes aka Erich Halk) tells this to

> us for the first time much earlier. In the August, 1952 issue of the magazine *Mensch und Schicksal*, he states that often in the last days of the war, in areas of highest priority for the Germans, the symbol of the "Schwarzen Sonne" (Black Sun) was in use in portions of the German armed air defense forces. These "heretics" within the SS, as Schweikhart calls them, above all used the air base at Banak in Northern Norway as their base for long distance flights over the inner Arctic using the most modern long distance aircraft which they had obtained after these aircraft were designated as "failed constructions" by the official Luftwaffe. Speaking specifically of this base at Banak, Schweikhart says: "If they likewise used a special symbol, it is uncertain, but possible." This "failed construction" methodology for procuring long-distance aircraft is repeated by Landig and said by him to include flying discs. These aircraft were the property of a sect within the SS and were re-branded with the Black Sun symbol according to Landig.
>
> Meanwhile, back in the highland plateau region, the Germans there evidently had ideas of carrying forth the war from Norway or at least prolonging the war from there. In fact, the last known German unit to surrender on the European mainland did so at Haudegen, Norway, on September 4, 1945. This date lacks four days from being four months after Admiral Doenitz officially surrendered the three German fighting divisions of the German military, the Army, the Navy and the Air Force. Please note that the SS apparently never officially surrendered.[46]

Stevens tells us that most of the information we have about these secret Arctic bases manned by the SS comes from Wilhelm Landig's book published in the 1970s called *Goetzen Gegen Thule* (*Idols Against Thule*), the first book in a trilogy (we explored the second book, *Wolfszeit um Thule*, above). Early in this novel three SS officers who are stationed in northern Norway leave their base in a special long-distance aircraft bound for a secret German postwar base somewhere in the Arctic, "Stuetzpunkt 103," or as it is commonly called, Point 103. Says Stevens:

> Wihelm Landig is just about our only source for the secret base he calls Point 103. Landig uses every setting, every situation, to unveil something that happened during the war or some aspect of thinking or technology. When his three characters, the three German soldiers, land at Point 103 they are told of the "V-7", a German flying disc that has been brought there. They are told its developmental history and means of operation. They are likewise informed of other prototype aircraft that have been brought to this base and housed there as well as the extent and purpose of Point 103 itself.
>
> Point 103 is set by Landig to be in the Canadian Arctic. We know that the Germans had a base as far West as Cape Chidley which is east of Hudson Bay but Landig's locating this particular base in the Canadian Arctic rather than farther east may be literary license, we simply do not know. It is set on an island in a semi-circular basin surrounding Point 103. Tunnels were made in the rock and served as living areas and storage as was done by the Germans at many other of their secret facilities, Nordhausen, Jonastal, Der Riese, Zement, and so on. Since aircraft were flown into this base, it must have had runway facilities. Hangars were blasted out of the solid rock mountains surrounding the base and it was in these cave-hangars that experimental aircraft of all sorts were stored. This base contained workshops, research facilities, housing, cooking facilities, offices, corridors, and assembly halls. Hundreds of German soldiers inhabited this base on an ongoing basis. It functioned as a listening station, air base, weather station and most probably a U-boat facility.

Stevens says that Point 103 was connected to the outside world with special radio transmissions that were sent to a special radio base in the Northern Andes. This may be the radio station that the Germans allegedly built at Machu Picchu near Cuzco, Peru. The Germans also built the beer brewery in Cuzco during this same time and that brewery is one of the few industries that exist in Cuzco. Stevens also says that flying disks were used to fly to various parts of the world to remain in contact with Nazi elements that remained after the war. Says Stevens:

> Writer O. Bergmann states that Germans of the Arctic groups were in communication with other post-war German groups via in intermediate position in the Northern Andes. Each group had its own code-name. The group in the Northern Andes used the designation "Atlantis," for instance. He states that other post-war groups survived in bases in inner Asia, as well as Africa and that these along with the Arctic base formed a unit while another three group unit was formed by a group in Greenland, somewhere near Australia and somewhere in the Pacific.

Stevens says that some of the craft kept at Point 103 were part of a special SS squadron and that these craft began using the Black Sun symbol as their insignia, rather than standard German insignias such as the Iron Cross. Says Stevens:

> It seems that all branches of the German military were represented at Point 103, including the SS. The particular sub-group or inner circle of the SS present at Point 103 lent their esoteric symbol, the Black Sun, to the base as a whole. On their aircraft the standard German insignias were removed and in their place the Black Sun sign was displayed. This took the form of a deep, dark-red circle. It is quite obvious that the thinking behind this symbol came from the Thule Society and their belief in a kind of Nordic Atlantis in the far north from which sprung the Aryan peoples, particularly the ancient Germanics. The Black Sun symbol reached its zenith at Point 103. It was adopted by some mystery U-boats of German origin sighted after the war but it was apparently not used in Neuschwabenland, Landig's Point 211. The flying discs and possibly aircraft flying from Point 211 used the old German military insignias.[46]

Stevens mentions how during the war, Wilhelm Landig had been in charge of security for the airport at Prague where three flying disc projects were housed. Since this area was the industrial heartland for the SS, Landig would have also been familiar with any aircraft projects under development by the SS. Stevens feels that Landig is a good source of information. Landig goes on to describe some other amazing aspects of alleged Nazi technology under development or operating at the workshops located at Point 103.

Landig says aircraft were guided to Point 103 by a special compass, called a Himmelskompass (heavenly compass). This compass did not work on magnetic lines of force generated by earth's magnetic field as most compasses do. This is because the magnetic north pole is not located at the geographic North Pole, but instead on the Boothia Peninsula near Hudson Bay, many hundreds of miles south of the geographic North Pole. Instead of measuring magnetic lines of force with a needle, the Himmelskompass measured polarized light. Using this device as a guide, an aircraft could be navigated to the proper coordinates without the use of a traditional compass and without interference from a defensive device employed at Point 103, the Magnetofunk. Says Stevens:

> The Magnetofunk was a compass in reverse. It generated a magnetic field and so would confuse any aircraft using a magnetic compass. It would influence the compass just enough to throw off anyone following it as a guide. Therefore, enemy aircraft would be passively diverted around Point 103 without the searching aircraft even being aware of its influence. ...
>
> But Landig describes other amazing devices and research going on deep within the bowels of Point 103. He states that a kind of alchemy is at work here. This is not the old alchemy of gold making, conspiracy or black cats, but a new alchemy for modern times. Vril energy was being analyzed. Landig goes on to state that this Vril energy was suitable for powering aircraft. It was the same sort of energy sought by the ancient Indians for their flying craft, the Vimanas. Mercury was being used as a partial means of propulsion. Landig then goes on to make an amazing statement, saying that "our Indian friends" are busy reconstructing ancient secrets under security observation from these ancient

sources.[46]

The Mysterious "Blue Island" of the Arctic

Associated with Point 103 is a "Blue Island" that was seen by Canadian air patrols, but was difficult to find when the air patrols returned to the same area for futher looks at this Arctic anomaly. Stevens says that routine Canadian air patrols originating out of the Parry Islands began encountering something they could not explain. Says Stevens:

> Several independent reports from around 1950 state more or less the same thing. It seems that these patrol flights (their purpose was not stated) would sometimes encounter what they described as an island, surround by an icy ring of high mountains. On this island building structures could just be made out. Further, aircraft of some sort which could not be properly identified were seen. These "notable powered aircraft" could be seen landing.
>
> But there was a big problem. All attempts to approach this island for a better look were met with a strange kind of resistance. Upon drawing closer to the island, the aircraft was surrounded by a "thick blue aether" which was impenetrable to the aircraft's radar.

Stevens points out that these Canadian flights were looking for UFOs and their base and the Americans were making similar flights out of Alaska. The famous UFO investigator Colonel Wendelle Stevens said numerous times that he was part of these missions in the years right after the end of WWII. With the stories of the Blue Island it would seem that some defensive device was being employed that hid the location of the secret base.

Other evidence of a German base in the Canadian arctic is the movement of the famous German U-boat, U-234, which surrendered near Nova Scotia at the end of the war. The U-234 was a Type Xb minelaying submarine. Type X U-boats were the largest submarines built by the Third Reich, displacing 2,000 tons. In 1944 the U-234 was refitted from a minelayer to a transport ship. It was headed for Japan at the very close of the war, but somehow ended up surrendering to Allied forces between Halifax, Nova Scotia and Portsmouth, New Hampshire, on May 14, nearly ten days after Germany had surrendered. The U-234 departed Kristiansand, Norway for Japan on April 15, 1945 but ended up surrendering a month later near Nova Scotia. What was the U-234 doing in Canadian waters when its mission had supposedly been to take cargo to Japan? This craft should have been in the South Atlantic. Had it come from the mysterious Point 103, the Blue Island in the Canadian arctic? It would seem so.

Landig's secret base at Point 103 was said to be comprised of members of the Thule society, a society which was sometimes at odds with the traditional power of the Nazi party and the Third Reich. They believed that Hitler and the Third Reich had been compromised by evil forces as early as 1933. Therefore, Point 103 was comprised of mystics and scientists from all over the world. In *Goetzen Gegen Thule,* Landig describes a conference at the base which is attended by the following:

> …a Tibetan lama, Japanese, Chinese, and American officers, Indians, a Black Ethiopian, Arabs, Persians, a Brazilian officer, a Venezuelan, a Siamese, and a full-blooded Mexican Indian.

Landig says that Point 103 is not an extension of the Third Reich, but a Thulean independent group opposed to fascism, working with esoteric groups from around the world that are in opposition to the Judeo-Christian/Masonic World Order. The focus and symbol of this group is the Black Sun insignia.

Landig describes the large gathering of esoteric groups from around the world that dedicated themselves to the Black Sun cause; they wore either red robes or black robes, black robes being for the higher initiates. He said the scene was obscured by a curtain, but behind the curtain when it opened up was a spacious hall which bore a set of steps leading downward. On both sides of the elongated hall ran some benches on which some of the men of the station sat. The aisles were lower and the processional path of the hall continued through four more levels. Upon this path stood a procession of people that for the most part wore red capes. Says Landig:

> At the head of the procession were foreign guests, whose attire also emphasized the strangeness of this assembly. Over everyone dominated the black, helmet-like headdress of the Tibetan Ta Lama, the Japanese next to him becoming small. While the Tibetan wore the already-known black robe, the Japanese officers had their uniforms on as well, however, also having black robes draped around them.

We have quite a curious group that assembled at this secret Arctic base at the end of WWII. We can see how there might be Japanese officers at the base, but the other people are incongruous, especially the Tibetan Lama. A group of Tibetan Lamas was discovered in Berlin at the end of the war and there are rumors of a secret Nazi base in Tibet. We will discuss both of these subjects in the next chapter. Also notable is that the Black Sun insignia is also a yantra—a meditation device to focus the eyes on—and its use is common in Tibet and India.

Greenland and the Lab of Death

The secret base in Greenland with the code name "Beaver Dam" is also associated with the so-called Laboratory of Death. Stevens said that he first heard about it at a UFO conference, and thought that it was located in Norway somewhere. He later discovered that most of the stories put the secret laboratory in Greenland, and that the rumors were that strange experiments went on there, including the grafting of different animal organs and limbs. Another version, he says, is that this fusion is done on a genetic level, rather than surgical. Obviously the whole idea is horribly creepy and one hopes that there is no such activity.

Stevens says that the fundamental reference concerning the Laboratory of Death is Claude Schweikhart and his 1952 article in the German periodical *Mensch und Schicksal*. He says that the article does not mention genetic manipulation or cloning or chimeras. Schweikhart reminds us of the investigation of the SS undertaken by the Louis Emerich team in conjunction with the Yugoslavian intelligence agency. This is the group that coined the word "Dritte Kraft" (Third Power) for the SS survival groups. Emerich and the

Yugoslavians concluded that a base did exist somewhere in Greenland which they called "Laboratoriums des Todes" (Laboratory of Death) but this base remains a complete mystery. Emerich and his group connect the secret base and laboratory with satanic activity.

But the big news from Schweikhart's article is that at the Laboratory of Death in Greenland developed a special type of flying craft—cigar-shaped or as a flying saucer—in 1951 (others had been developed earlier). This new flying craft had the ability to submerge in the ocean and emerge from the ocean to take flight into the air. In other words, it was a flying submarine. Says Stevens, addressing the facilities needed to develope such a thing:

> Schweikhart was writing in 1952 so when he says this development took place in an unrepentant Nazi secret base on Greenland in 1951, we can only assume Schweikhart was well connected to say the least. I do not expect any reader to take this statement, a statement from many years ago, from a mysterious source, describing an incredible event, at face value. But I would ask the reader to suspend judgment of the combination flying saucer-submarine until we can deal with that subject in some detail.
>
> So picture, if you will, a base capable of such things. This base would have to be large indeed to accommodate both U-boats and flying discs. Bergmann describes what a Greenland base must have looked like as being essentially the same type of base needed at the South Pole at Neuschwabenland. It would consist of tunnels bored through ice and solid rock to a length of 2,000 meters. The entrance would have been camouflaged so as to make it unfindable.
>
> According to Bergmann, this type of base would have been protected by two kinds of weapons, both called "Strahlenwaffen" or ray weapons. First, there was the weapon he calls "Zuendunterbrechung" or ignition interruption in English. This word is simply a variant of "Motorstoppmittel," called magnetic wave by the Americans. This was a device that ionized the air within its range of operation and so short-circuited the ignition systems of enemy aircraft. The second weapon "Strahlenwaffe" goes undescribed by Bergmann

> but we may conclude it was a more directed beam weapon since Bergmann goes to pains to describe the result of the magnetic wave-type weapon. We will encounter both types of weapons again.
>
> To flesh out our conception of such a base, we must assume parallel tunnels that are interconnected as they were in the underground facilities in Germany, Poland and the Czech Republic. Storerooms, living quarters, power generating facilities, heating facilities, and workshops would also be necessary. In addition, some sort of turntable for incoming U-boats would have been necessary in order to service them and then spin them around for launching again seaward. Access to a rough airfield would also probably be necessary for supplies and reinforcements from Norway.

Stevens thinks that "Beaver Dam" was a submarine base deep inside a hollow cavity in solid rock that was partially above water and partially below water:

> In the absolute maximum of possibilities for Beaver Dam, I would picture it as breakwater or near breakwater, allowing only the tops of the waves into a central harbor. The central harbor would remain relatively calm for a U-boat, especially if it were submerged. To access the base itself the U-boat would dip a few feet underwater under an artificial ledge built into solid rock. Behind the ledge the U-boat would surface into a large hollow cavity completely cut off from the surrounding environment. The cavity would contain lights, dry-docks, machinery for on and off loading.... The artificial breakwater and the reinforced cement ledge would be camouflaged to look like real rock or real rock might even be affixed on to it for this natural look. To aerial reconnaissance or even ships sailing past it, the base would blend into the natural surroundings. A small airfield was probably nearby since contact with Norway via air was necessary at least in the beginning.

So, starting in the early 1950s, this facility in Greenland—

run by the SS—developed a flying U-boat. These flying disks and cylinders are essentially submarines and airships, both UFOs and USOs. Some researchers even postulate the origin of the UFO that crashed at Roswell as coming from this base in Greenland. This is in agreement with the central thesis put forth by Dr. Joseph P. Farrell in his book *Roswell and the Reich*.[39]

Stevens tells us that other weapons were kept at Beaver Dam according to Dr. Milos Jesensky and Mr. Robert Lesniakiewicz in the 1998 German language book *Wunderland.* Among these weapons was the "Urzel." The Urzel was a combined V-1, V-2, mounted in a horse and rider configuration. This weapon was to be launched at the east coast of the USA and was to have contained an atomic warhead. Unfortunately for the Nazis, the atomic production facilities at Der Riese either simply could not produce the desired weapon in time or were overrun by the Soviets before they could do so.

Stevens says that Dr. Jesensky and Mr. Lesniakiewicz say that the mission and scope of Beaver Dam changed somewhat corresponding to the rapidly changing war situation. They list three stages for Beaver Dam:

> In stage one, Beaver Dam would have been the impregnable fortress Dönitz promised for Hitler. Hitler was to be able to carry out war outside of Germany in this vision so U-boats and even tug boats would be necessary since containers carrying large rockets capable of hitting the United States would have been brought in. This base also functioned as a meteorological base until the bitter end of the war in Europe. Training for the men as well as testing and working out of details involving Arctic living would have been carried out at the polar research station at the Golden Peak, Kirkonoshe, in the Bohemian mountains.
>
> In stage two the atomic warheads and very long range rockets could no longer be counted upon since the research, development and industrial facilities lying underground in the Owl Mountains of Poland were unable to provide them in a timely manner. Therefore, Beaver Dam took on a secondary function. It became a repository. Art, artifacts,

> gold and money from the Reich's Bank, blueprints for superweapons as well as prototype aircraft were brought to Beaver Dam for storage. Among these prototypes/superweapon blueprints was a flying disc.
>
> In stage three the war is over. The base now served as a hide-away for prominent Nazis on the run but the most important function of Beaver Dam centered around the flying discs. This was their base and point of origin for spying missions directed at stealing atomic weapons and technology from the USA and USSR. Urzel weapons were also in storage. The Nazis at Beaver Dam still prepared for some sort of final battle, a Goetterdammerung, in which they would fight with advanced weapons and machines of all kinds.[46]

Stevens says that these researchers believed that the Germans at Beaver Dam in Greenland wanted to cause provocation between the East and West. Reinhard Gehlen, who ran a German intelligence network inside the Soviet Union during the war and was recruited by the CIA after the war to reconstitute his SS spy network, is often accused of playing one side against the other, even manufacturing the Cold War, for the benefit of those he served in the Nazi period and their ideals. For the SS, the communist Russians were the enemy and the Americans were potential allies.

Stevens says that the Slavic researchers do not think Beaver Dam exists today. He comments:

> At some point during the last 60 plus years they believe the base ceased to function. They really do not say what might have happened but they do speculate about walking through the underground tunnels and finding abandoned, rusty flying discs and derelict U-boats. These researchers believe the evidence is there, only awaiting discovery. Imagine if it were true, if someone did find a secret German base, accessible only by tunnel to the sea, filled with U-boats and supporting machinery that had been in operation after 1945? That would settle the question of post-war German bases once and for all, right?

Stevens says the German researcher O. Bergmann, writing in 1989, says that just as Admiral Byrd led an expedition against the German base in the Antarctic, at Neuschwabenland, so too were expeditions mounted in Greenland, which Bergmann hints were for the same purpose. During the years 1947 to 1950 a Danish expedition commanded by Eigil Knuths was sent to Greenland. In the years 1947 to 1951 the French sent an expedition. In 1951, the English got into the act with their own expedition to Greenland. Says Stevens:

> But what is most amazing is that on March 27, 1951 the Danish, who own Greenland, found it necessary to enter into an agreement with the United States for the common defense of Greenland. Who were they defending Greenland against? Greenland is a sparsely populated land with no large cities, mostly inhabited by Eskimos still practicing variations of a hunter-gatherer lifestyle. The only military power in 1951 set up in opposition to the United States was the Soviet Union and it hardly had its sights set on invading Greenland.
>
> Immediately, the US government began feverishly building some air bases in Greenland. Bergmann presents a map of Greenland showing three bases, Narsarssuak, Thule and Station North. The base at Station North is located exactly where Dr. Solchec, Dr. Jesensky and Mr. Robert Lesniakiewicz think Beaver Dam was located, on the northeast peninsula called Peary's Land.
>
> So what happened here? The Germans were said to have just perfected their flying submarine at this time. How could the Americans and presumably NATO have a base in this exact spot or within a few miles of it?

Stevens says that a likely scenario is that the Germans were displaced sometime in the years 1951 or 1952. They could have abandoned this base or these bases or they could have been forced out by the American military. They might even have entered into an agreement to cede the bases. If this is true, the March 27, 1951 agreement may have involved a secret party, one representing the interests of the Nazis. The hidden base itself with its hidden access to the sea may have been appropriated as a listening post and for use in clandestine activities directed at the Soviet Union.

So we have explored the origins of the Black Fleet and the northern bases that were used by the dozens of submarines that continued to patrol the Atlantic and elsewhere. The bases in Greenland and the Canadian arctic apparently existed until 1952. No matter when these bases were closed or were taken over by the Allies, the activity continued in South America and Antarctica. Submarines—and even flying disks—continued to ferry people and technology from Europe, Spanish Sahara and the Canary Islands to Antarctica and South America. Other flights, as we will examine in the next chapter, were to Tibet.

Alleged to be an actual photograph of a Haunebu II on the ground.

Chapter Six

A Secret Base in Tibet

As we acquire knowledge, things do not become more comprehensible, but more mysterious.
—Will Durant

Before moving on to the obvious places for the Black Fleet and remnants of the Third Reich to have retreated to their refuges in South America and Antarctica—let us look at some of the long-range German aircraft and some of the other locations to which these new planes might have flown. Both Henry Stevens and Joseph Farrell suggest that the Germans established a base in Tibet during the war and used it as a repository for technology and artifacts.

Stevens says that by the term repositories, he means places where the Germans, particularly the SS, stashed high-value things for safekeeping after the war. These high-value things could be aircraft or flying discs or blueprints, artifacts of value in Nazi ideology, money, or even printing presses for making money. We have seen some alarming testimony about the laboratories and manufacturing facilities that were supposedly kept in Greenland for some years after the war had officially ended. Was there a secret base in Tibet as well? This base could not be reached by submarine and would need to be reached by long-range aircraft, or even via flying disk.

A Secret German Base in Tibet?

As Stevens says, "There is really no denying something going on between the Tibetans and the Nazis during the war."[46] He goes on to say that we do not know how many expeditions were made to Tibet by the Germans, but we do know about the 1938-39 SS

The Thirteenth Dalai Lama (1876-1933) in an official portrait in 1932.

Expedition to Tibet led by German zoologist and SS officer Ernst Schäfer.

The Reichsführer-SS Heinrich Himmler wanted to use the reputation of Schäfer for Nazi propaganda and asked the zoologist about his future plans in 1937. Schäfer responded that he wanted to lead another expedition to Tibet (he had entered Tibet from China some years earlier). Himmler told the zoologist that he would like to fund the expedition and make it an official undertaking of the SS Ahnenerbe (SS Ancestral Heritage Society).

British writer Christopher Hale, in his book *Himmler's Crusade: The Nazi Expedition to Find the Origins of the Aryan Race*,[37] says that Himmler was fascinated by Asian mysticism and therefore wished to send such an expedition under the auspices of the SS Ahnenerbe. Also, Himmler desired that Schäfer perform research based on Hanns Hörbiger's pseudoscientific theory of "Glacial Cosmogony" promoted by the Ahnenerbe. However, Schäfer had scientific objectives and therefore refused to include Edmund Kiss, an adept of Hörbiger's theory, in his team and required 12 conditions to ensure scientific freedom.

Himmler was agreeable to the expedition going ahead provided all members joined the SS including Schäfer. The expedition is widely known as the SS Expedition to Tibet 1938-39.

Christopher Hale observes that "while the idea of 'Nazi botany' or 'Nazi ornithology' is probably absurd, other sciences are not so innocent—and Schäfer's small expedition represented a cross-section of German science in the 1930s." To Hale, this has considerable significance as "under the Third Reich anthropology and medicine were cold-bloodedly exploited to support and enact a murderous creed."[37]

There are strong allegations, though never officially acknowledged, that one of the expedition's purposes was to determine whether Tibet was the cradle of the Aryan race. This was largely done by the taking of cranial measurements and the making of facial casts of local people by the anthropologist Bruno Beger.

Hale relates the existence of a secret warning issued by propaganda minister Joseph Goebbels to German newspapers in 1940 saying that "the chief task of the Tibet expedition," was "of a political and military nature" and "had not so much to do with the

The 13th Dalai Lama of Tibet, Sir Charles Bell (both seated), and Sidkeong Tulku Namgyal of Sikkim in 1910.

solution of scientific questions," adding that details could not be revealed.[37]

Obviously, the expedition was caught up in the politics of its time, and whenever the SS was involved there was likely some military and technological purpose to the expedition as well. It was suspected at the time that the Germans were scouting out a possible location for an airbase located somewhere on the plateau of Tibet. They could use this airbase to attack the British army in India, which was otherwise beyond the scope of the German air force.

According to Wikipedia:

> Chinese journalist Ren Yanshi, quoting the Austrian weekly *Wochenpresse*, writes that the first major task of the expedition was "to investigate the possibility of establishing the region as a base for attacking the British troops stationed in India" while its second major assignment was "to verify Heinrich Himmler's Nazi racial theory that a group of pure-

blooded Aryans had settled in Tibet."

According to American journalist Karl E. Meyer, one of the expedition's aims was to prepare maps and survey passes "for possible use of Tibet as a staging ground for guerrilla assaults on British India."

After traveling by boat from Europe to Calcutta, the SS expedition team assembled in Sikkim's capital of Gangtok, at that time a British protected mini-state like Bhutan. The expedition assembled a 50-mule caravan and searched for porters and Tibetan interpreters. Hale reports that at Gangtok, the British official, Sir Basil Gould, observed them, describing Schäfer as "interesting, forceful, volatile, scholarly, vain to the point of childishness, disregardful of social convention," and noted that he was determined to enter Tibet regardless of whether he got permission.[37]

The team began their Tibetan journey on June 21, 1938, traveling as a caravan through the Teesta River valley and then heading north. In August 1938, a high official of the Rajah Tering, a member of the Sikkimese royal family living in Tibet, entered the team's camp. Schäfer met with the official, and presented him with mule-loads of gifts.

Expedition members with hosts in Gangtok, Sikkim are (from left to right) unknown, unknown Tibetan, Bruno Beger, Ernst Schäfer, Sir Basil Gould, Krause, unknown Tibetan, Karl Wienert, Edmund Geer, unknown, unknown. 1938.

In December 1938 the Tibetan council of ministers in Lhasa formally invited Schäfer and his team to Tibet, but they were forbidden to kill any animals during their stay, due to religious concerns. After a supply trip back to Gangtok, Schäfer learned he had been promoted to SS-Hauptsturmführer, and the rest of the team had been promoted to SS-Obersturmführer because Himmler was so excited about the expedition having successfully entered Tibet.

The team reached the capital of Lhasa on January 19, 1939. Schäfer proceeded to pay his respects and offer gifts to the Tibetan ministers and a nobleman. He also gave out Nazi swastika pennants, explaining the reverence shown for the shared symbol in Germany. They demonstrated their high-quality German rifles and pistols to the Tibetans. The expedition's permission to remain in Lhasa was extended, and Schäfer was permitted to photograph and film the region. The team spent two months in Lhasa and the immediate area, collecting information on the culture, agriculture, and religion. One might also think that they were looking for a suitable place to build an airfield, and a hangar that was inside a cliff or mountain.

Schäfer met the Regent of Tibet, Reting Rinpoche, on several occasions. During one of their meetings, the Regent asked him point

Under SS pennants and a swastika, the expedition members are entertaining some Tibetan dignitaries and the Chinese representative in Lhasa; left: unknown, Beger, Chang Wei-pei Geer, unknown, unknown; in the center: Tsarong Dzasa, Schäfer; right: unknown, Wienert, Möndro.

blank whether his country would be willing to sell weapons to Tibet.

In March 1939, the expedition left Lhasa, heading for Gyantse escorted by a Tibetan official. After exploring the ruins of the ancient deserted capital city of Jalung Phodrang, they reached Shigatse, the city of the Panchen lamas, in April. They received a warm welcome from the locals, with thousands coming out to greet them. Here, Schäfer claims to have met "the pro-German regent of Shigatse" (the 9th Panchen Lama had died in 1937 and the 10th was not to arrive before 1951). All in all the Germans were made to feel welcome in Tibet and the Tibetans seemed to like the idea of being allies with the Germans, who might aid them in their efforts to keep the British and Chinese from taking too much control of the country, especially in terms of foreign relations with outside countries. As mentioned previously, the Reting Rinpoche had asked if they could buy weapons from the Germans. In fact, these weapons would probably just be given to the Tibetans, along with a German SS unit to administer the airbase and arms depot.

In May of 1938, the expedition returned to Gyantse and Lhasa and began negotiations with local British officials about the trip back to India and transport of the expedition's gear and collections. After Schäfer read a letter from his father who reported to him about the imminent threat of war, and urged him to return to Germany as quickly as possible, Schäfer decided to depart immediately from Lhasa.[37]

After being given two complimentary letters, one to Hitler and the other to Himmler, Schäfer and his companions left. They also took with them two presents for Hitler, those being a Lama dress and a Tibetan hunting dog, as well as a copy of the Tibetan "Bible," the 120-volume *Kang Shur*. The expedition headed south over the Himalayas to Darjeeling and arrived in Calcutta in July of 1939. After a brief stay the Germans boarded a British Airways seaplane at the mouth of the Hooghly River, and began the journey home via Karachi and Baghdad.

Schäfer kept meticulous notes on the religious and cultural customs of the Tibetans, from their various colorful Buddhist festivals to Tibetan attitudes towards marriage, rape, menstruation, childbirth, homosexuality and masturbation. In his account of Tibetan homosexuality he describes the various positions taken by older lamas with younger boys and then goes on to explain how homosexuality played an important role in the higher politics of Tibet. There are pages of careful observation of Himalayan people engaged in a variety of intimate acts. Schäfer presented the results of the expedition on July 25, 1939 at the Himalaya Club Calcutta.

Throughout his stay in Lhasa, Ernst Schäfer had remained in touch with Germany through mail and the Chinese Legation's radio. Heinrich Himmler was reported to have followed the expedition very enthusiastically, writing several letters to Schäfer and even broadcasting a Christmas greeting to him via shortwave. One must presume that, no matter how scientifically pure Schäfer's motives were, the SS would have put several special officers, perhaps unknown to Schäfer, into the team to look for possible airfields and hangars.

Tibetans in Berlin

According to a number of sources, on April 25, 1945, a group of Russian soldiers entering Berlin discovered the dead bodies of six Tibetan monks in a circle, with the body of a seventh monk in the center. The dead man in the center was wearing a pair of bright green gloves. One of the Russian soldiers was a Mongolian and he literally freaked out, dropping to his knees and praying to, or perhaps for, the dead monks. Over the next few weeks the bodies of hundreds of Tibetan monks were found in the ruins of Berlin; all were said to have committed ritual suicide.

What were these monks doing in Berlin and how did they get there? It would have been difficult for these Tibetans to have come to Germany by sea as they would have had to travel through British India and take a ship from a British-held port. Nor could these Tibetans have come overland through China or Russia (via Mongolia) as these countries were at war (with Germany, in Russia's case). So it has been theorized that these Tibetan lamas were flown from Tibet to Germany with all of their religious artifacts and costumes.

This would mean that when the SS returned to Germany they had made plans to make air contact with Tibet and build an airbase on the Tibetan plateau somewhere. I would suggest in western Tibet, possibly even near Mount Kailash, or further west, where the land can be flat and dry with snowy peaks surrounding large, flat valleys. There are many places in western Tibet where only minimal work would need to be done to create a long flat landing spot where a large airplane could safely land. The distance from this area to Austria is approximately the same as from France to New York City, and these flights were meant to be non-fueled return flights. Therefore, a trip to Tibet from Austria would have been a round-trip event with no need for refueling at the new airbase in Tibet.

And the flight could have contained barrels of fuel to be stockpiled for the future need to refuel an airplane while at the airbase in Tibet. Future flights could have continued to add to this cache until a considerable stockpile of fuel (and spare parts) existed in this remote part of Tibet. It seems probable that a stockpile was accumulated that could have refueled a number of these thirsty aircraft. However, such a stockpile would not last forever, nor would an isolated base in Tibet.

Dr. Joseph P. Farrell gives us a rather rational description of what might be considered a German base in Tibet in his 2008 book *Nazi International.*[12] Says Farrell:

> The likely possibility was sometime around the period of the massive German offensive on the southern Russian Front in the summer of 1942. It is a little known fact that the Wehrmacht sent armed long-range reconnaissance units as far east as Astrakhan on the northern Caspian Sea, but beyond sowing confusion behind Russian lines, there was little military value in such an operation. [Here Farrell cites the German military historian Paul Carrell in the book *Hitler Moves East: 1941-1943*] However, it is possible that commando-engineered units of the Waffen SS might have been included in these units, and, once the units had penetrated as far east as was operationally safe, the commando units would have been released to continue onward through Kazakstan and on into Tibet, utilizing Nazi intelligence contacts in these Muslim areas to smooth their passage through hostile territory. Once in Tibet, a makeshift airfield could have been constructed able to handle long-range aircraft. The Tibetans would have then been flown back to Germany.

> While a hazardous enterprise to be sure, it would have been far safer than attempting to smuggle monks out through British India and back to Germany via U-boat, which, in any case, would have required more trips than would have been required by even one of the large Junkers 290s or the enormous Junkers 390s. As for the purpose of such an operation, this is relatively transparent, for such expertise would have been required if Himmler's SS was to translate its copy of the *Kang Shur* completely and accurately.
>
> This requires some commentary. Since the 1938-1939 Schaefer Expedition is known not to have brought out large numbers of Tibetan monks, but only a copy of the *Kang Shur*, the question becomes, when were these monks smuggled out of Tibet into Nazi Germany, by what route, and for what purpose?[12]

Yes, this all very possible. Farrell seems to think that the SS might have been able to continue overland from Kazakstan to Tibet to establish an airfield, but it seems more likely that any travel into Tibet was by aircraft, and Kazakstan may have been a good location to launch the initial flights into Tibet. Weapons plus other supplies for a base were flown into Tibet, I suggest western Tibet, and a makeshift airfield was created for long-range aircraft.

What I propose happened is this: after a weapons deal with Tibetan ministers in Lhasa was concluded, several SS members of the SS Expedition stayed behind, or had secretly met other SS officers in Lhasa who had come via China.

These SS officers that were left behind in Lhasa were then given a small Tibetan unit, and along with a transmitter that they brought with them from Germany, went to western Tibet to find a suitable airbase for a long-range German aircraft to land. This small team of perhaps two Germans and ten Tibetans could have set up an airfield and beacon that would allow these long-range aircraft to fly from Austria all the way to Tibet, carrying fuel, weapons, generators, lights, and more SS officers. Prior to 1944, when the very long-range aircraft first became available, they would have had to fly out of airbases in the captured sections of the Ukraine, much closer to

Tibet than Austria, or carry extra drums of fuel as well as make a stopover somewhere in Turkey, Iran, or Afghanistan (such as the western city of Herat).

After several such early flights bringing machinery and supplies, a fairly large number of Tibetan lamas could have been brought to Germany for special use by the SS. Translating the *Kang Shuur* texts into German would have been one of their duties. Supplying the Tibetans with brand new German rifles of even the most basic type would fulfill Germany's "promise" to arm Tibet. All of this was happening about 15 years before the Dalai Lama, as a young boy, was to flee Tibet as the Chinese took over the country.

A curious tale is told by the American doctor Howard Buechner, who had co-authored with Bernhart on the U-530 books, in his 1991 book *Emerald Cup—Ark of Gold*,[30] of a flight to Tibet from Austria late in the war. But first some background information.

In the summer of 1931 a German mystic and SS officer named Otto Rahn went to the Languedoc area of southern France on a secret mission for Heinrich Himmler and the Nazi SS. His destination was the Cathar fortress of Montségur and his mission was to find the lost treasure from Solomon's Temple, including the Holy Grail as an emerald-encased gold cup. Otto Rahn was said to have discovered a cavern system beneath Montségur that led him to discover (apparently) a portion of this treasure. Otto Rahn was a fascinating person, an officer who was a scholar and a mystic. He believed in Atlantis, reincarnation, and that he had been a Cathar in a previous life.

According to former US Army colonel Howard Buechner in *Emerald Cup—Ark of Gold*,[50] in February of 1944 the famous SS commando Otto Skorzeny recovered the treasure from the caverns beneath Montségur. Otto Rahn had died mysteriously the year before, apparently assassinated from within the Third Reich. According to Buechner, the treasure that was so carefully guarded for thousands of years eventually went to the Berchtesgaden fortress high in the Bavarian Alps, Germany's famed Eagle's Nest. In a scene worthy of the Indiana Jones film *Raiders of the Lost Ark*, the treasure was taken aboard a Nazi transport plane and eventually flown to a secret destination.

Says Buechner (apparently with information Bernhart):

> In the very last days of April, 1945, eyewitnesses noted the mysterious takeoff, in the region of Salzburg, Austria, of a four engine aircraft believed to be a Heinkel 277 V-1 [Version-1]. The destination of the plane has remained unknown, but some authors have proposed that it flew to the city of Katmandu in Nepal and then to some other location in the Himalayas or in Tibet. On board were five officers of the Black Order and in the cargo bay were the twelve stone tablets of the Germanic Grail which contained the key to ultimate knowledge.[50]

Buechner believes that the Holy Grail and the Ark of the Covenant, as well as other treasures, were taken to a secret hideaway somewhere in Tibet. Buechner is apparently referring to German authors, writing in German, when he says, "some authors."

This may seem far-fetched, but there is a great deal of literature supporting close associations between the Third Reich and certain remote areas of Tibet whose locations are unknown. Given the Nazis' obsession with the occult, a destination such as Kathmandu or Tibet would not seem out of place. The suggestion that the flight might have landed in Kathmandu, rather than having flown directly to Tibet, is an interesting one. At this time in history, Nepal was a neutral nation, one that largely forbade outsiders from entering the country, but had strong ties to both Tibet and India, and many ethnic Tibetans live in Nepal. Most of Tibet's bronze, silver and gold statuary was actually made in the Kathmandu Valley and exported to Tibet.

The Nepalese probably would have allowed a German plane to land at their small airport on the fringes of Kathmandu, near the Himalayas, Tibet and Mount Everest. However, it would seem that the German planes might just as well have flown directly to the airbase in western Tibet, which is actually closer to Austria than Kathmandu.

Buechner suggests that the craft flying out of this far-eastern part of the Third Reich originated in Austria and flew directly east over Romania, the Black Sea and the Caspian Sea and into western Tibet. This flight can easily be made today by modern jets but during

A Heinkel He-277 four-engine long-range aircraft.

World War II, until its last year or so, such a flight was virtually impossible. Planes could carry drums of fuel on them and make stopovers along the way. Airports in Iran or Afghanistan, during the war, would be possibilities for such a stopover, but as we shall see, they were unnecessary.

There were long-range aircraft available at the end of the war and one of them was the Heinkel 277. The Heinkel He-277 was a four-engine, long-range heavy bomber design, originating as a derivative of the He-177. The He-177 used two Daimler-Benz DB 606 "power system" engines, while the He-277 used four similar engines.

The Heinkel 277 had a crew of seven and flew at a maximum speed of 570 km/h at 5,700 meters (354 mph at 18,700 feet) with a cruise speed of 460 km/h (286 mph). Its range was up to 11,100 km (6,900 miles) for the Amerika Bomber version that was to be able to bomb New York City and return without refueling. This plane was Heinkel's entry in the important transoceanic range Amerikabomber competition with other aircraft manufacturers during the war to make long-distance flights, such as from France to New York City and back, without refueling in order to make long-range bombing missions against American targets.

However, it is said that the design was never produced and no prototype airframe was completed. Supposedly, the deteriorating condition of the German aviation industry late in the war and the competition from other long-range bomber designs from other firms, led to the design being cancelled. However, photos of completed He-277's are known, and shown here, and it is likely that at least four or five He-277s were made in the early process of converting the He-177 into a four-engine long-range bomber or cargo plane.

This was typical during World War II on all sides; the boys

A Messerschmitt Me-264 V-1 four-engine long-range aircraft.

in the field did what they could and flew what they could, even if they were prototypes that did not fulfill all of the specifications of the final aircraft that was to the be the He-277. A number of these prototypes were undoubtedly built as the major components and the airframe basically already existed. The big question is where did these missing long range planes fly to? Did one of them fly on a mission to Tibet at the end of WWII as Buechner and Barnhardt claimed? Where did the other He-277 four-engine airplanes depart to? Perhaps to Greenland, Spanish Sahara, the Canary Islands or even as far as Antarctica (probably with a stopover in a secret SS airfield in the Spanish Sahara)?

The Heinkel-277 was one of four different long-range bomber prototypes that the Nazis had commissioned from the active aircraft manufacturing industry in Germany, Poland and the Czech areas. The He-277 had huge tires, long wings, four engines and a big glass dome area for the pilots and crew to sit in at the front of the fuselage. It would have been an impressive sight at any airfield in Europe or America at the time, and certainly would have amazed people in remoter areas of the world, like Tibet, with its sheer size.

The Americans were doing much the same thing with the new aircraft that they had commissioned. If one industry was winning during WWII you could say it was the aerospace industry. Billions of dollars in wealth were poured into aerospace by all sides during this time. The Japanese were similarly developing their larger aircraft as well.

The Germans had four different companies creating large, long-range aircraft to continue the war on a larger scale. They were frustrated in their efforts to transfer technology to the Japanese via submarine and attempted flights over the pole to Japan, which were apparently successful.

Besides the He-277, there was the Messerschmitt Me-264, the Focke-Wulf Ta-400 and the Junkers Ju-390. All of these aircraft were long-range, four- or six-engine bombers (that could also be cargo planes), and the Third Reich pinned much of its effort to win the war on these new long-range strike aircraft. In fact, it was said by Albert Speer, that a secret Ju-390 flight to Japan had occurred "late in the war." This flight, by a Luftwaffe test pilot, was a non-stop flight, probably from northern Norway, to Japan via the polar route. While some historians doubt that this flight took place, it probably was made and maybe more that once.

The Junkers Ju-390 carried a crew of ten with six engines and a longer range than the He-277; it made its maiden flight on October 20, 1943. It performed well and this resulted in an order for 26 aircraft, to be named Ju-390 V1.

Two prototypes were known to have been created by attaching an extra pair of inner-wing segments onto the wings of basic Junkers Ju-90 and Ju-290 airframes and adding new sections to lengthen the fuselages. It is thought that at least three Ju-390 prototypes were built and possibly many more. After the war it was claimed that the SS commandeered these prototype planes and put them at various SS bases in northern Norway, Greenland, and elsewhere—quite possibly Tibet.

On June 29, 1944, the Luftwaffe Quartermaster General paid

A Messerschmitt Me-264 V-1 four-engine long-range aircraft in flight.

A Junkers Ju-390 six-engine long-range aircraft.

Junkers to complete seven Ju-390 aircraft. The contracts for the 26 Ju-390 V1s were cancelled. Instead, a Version 2 (V2) was to be commissioned. At this point there were at least two prototypes of Ju-390 V1, and possibly as many as seven. They probably also had several other long-range prototypes, such as the He-277 prototypes, and therefore the SS at the very end of the war may have had 12 or more long-range aircraft that could carry passengers, cargo, machine parts and even arms to places as far away as Japan, South Africa and South America.

It is thought that the Ju-390 V2 was completed within months with flight tests beginning at the end of September 1944. However, some state that the Ju-390 V2 was assembled in Bernburg, Germany and first flown in October 1943. This would place its construction and first flight at nearly the same time as that of the Ju-390 V1. After the war it was claimed that only one Ju-390 V2 was ever flown. Wikipedia says that at a hearing before British authorities on September 26, 1945, Professor Heinrich Hertel, chief designer and technical director of Junkers Aircraft & Motor Works asserted the Ju-390 V2 had never been completed. However, the site says that German author Friedrich Georg claimed that test pilot Oberleutnant Joachim Eisermann recorded in his logbook that he flew the V2 prototype (RC+DA) on February 9, 1945 at Rechlin air base. The log is said to have recorded a handling flight lasting 50 minutes and composed of circuits around Rechlin, while a second 20-minute flight was made to ferry the prototype to an airbase at Lärz.

So, it would seem that after the war the Germans tried to claim that only one test flight of one Ju-390 ever occurred, and that this single plane never flew on any missions. However, most researchers think that this is completely false, with as many as nine or 10 Junkers 390 (both versions) being built and flown. It would seem that the special SS squadrons obtained several, if not all, of the Ju-390 airplanes, and used them on secret missions.

Some of the secret missions that the Ju-390 allegedly went on

A Junkers Ju-390 six-engine long-range aircraft in flight.

were a flight to Cape Town in South Africa, a polar flight to Japan, a dry-run bombing mission on New York City, and possibly several flights to Tibet. It may have made flights after the war to Greenland, Spanish Sahara and South America. From Tibet flights could have been made to Indonesia or other parts of the Far East.

Wikipedia says this about the alleged flight to South Africa:

> A Ju-390 is claimed by some to have made a test flight from Germany to Cape Town in early 1944. The sole source for the story is a speculative article which appeared in the *Daily Telegraph* in 1969 titled "Lone Bomber Raid on New York Planned by Hitler," in which Hans Pancherz reportedly claimed to have made the flight. Author James P. Duffy has carried out extensive research into this claim, which has proved fruitless. Authors Kössler and Ott make no mention of this claim either, despite having interviewed Pancherz.

It should be remembered that during WWII the Nationalist Afrikaners were pro-Germany but the British put the country under martial law during the war to combat the pro-leanings of the many German and Afrikaner residents of South Africa and Namibia. Namibia, or possibly Angola near the Namibian border, would be another possible site for a secret Nazi airbase.

On the round-trip New York flight Wikipedia says:

The first public mention of an alleged flight of a Ju-390 to North America appeared in a letter published in the November 1955 issue of the British magazine *RAF Flying Review*, of which aviation writer William Green was an editor. The magazine's editors were skeptical of the claim, which asserted that two Ju-390s had made the flight and that it included a one-hour stay over New York City. In March 1956, the Review published a letter from an RAF officer which claimed to clarify the account. According to Green's reporting, in June 1944, Allied Intelligence had learned from prisoner interrogations that a Ju-390 had been delivered in January 1944 to Fernaufklärungsgruppe 5, based at Mont-de-Marsan near Bordeaux and that it had completed a 32-hour reconnaissance flight to within 19 kilometers (12 miles) of the U.S. coast, north of New York City.

This flight has been much discussed by Joseph Farrell in his many books, including *Reich of the Black Sun*.[30] In the book, Farrell says that the Allies were worried about an atom bomb attack by the Germans in 1944. He says that in 1945, Hitler insisted that holding Prague could win the war for the Third Reich because of new technology being developed there. He says that for this same reason US General George Patton's Third Army made a race for the Skoda works at Pilsen in Czechoslovakia instead of heading for Berlin. He also says that the US Army did not test the uranium atom bomb it dropped on Hiroshima because it came from the Germans. Farrell argues that Nazi Germany actually won the race for the atom bomb in late 1944, and then goes on to explore the even more secretive research that the Nazis were conducting into alternative physics,

The Japanese Nakajima G8N1 four-engine long-range aircraft.

A captured Japanese Nakajima G8N1 four-engine long-range aircraft.

new energy sources, and the occult. This is the first book in English to look at the Nazi Bell project and the Kecksburg, Pennsylvania UFO crash in the light of the super-secret black projects being run by the SS.[30]

On the intriguing polar flight to Japan Wikipedia says:

> In his book *The Bunker*, author James P. O'Donnell mentions a flight to Japan. O'Donnell claimed that Albert Speer, in an early 1970s telephone interview, stated that there had been a secret Ju-390 flight to Japan "late in the war." The flight, by a Luftwaffe test pilot, had supposedly been non-stop via the polar route.

So, it is an intriguing possibility that a number of long-range, newly built aircraft were making secret flights to various parts of the world from areas of Europe that the Third Reich still controlled at the end of the war. It makes perfect sense, and now that it is well established that a number of U-boats escaped into the Atlantic after the war, we can see that aircraft were used as well. The big question is where did these flights go and what was their cargo?

Like the U-boats at the end of the war, it was not necessary for these long-range bombers to go on bombing missions any more. These planes still had some weaponry, such as machine guns, but they would now be used for long distance cargo flights. The surviving Nazis, mainly SS officers, were entering a new phase and these aircraft were now passenger and cargo planes, carrying printing presses for printing money, escaped Nazis, gold bullion, technical plans, and other documents.

Atlantis, Tibet and Secret Science

With the strange depictions of Tibetan lamas at the secret submarine base called Point 103 in the Arctic we can see that the SS placed a great deal of importance on Tibet, including its location and history. Henry Stevens explains that the Nazis saw their origins in a Tibetan-Aryan culture of India, Tibet and the northern Baltic Regions. They also believed in Atlantis and in technical achievements made by civilizations in the remote past. Says Stevens:

> The problem was that European pre-history during the 1930s and before was centered about the Near East, so, allegedly, it was those people who were solely responsible for bringing "culture" to Europe. I say "Near East" but it took on tones of being bible-centric. The Nazis rejected this foreign origin of their culture and wanted to explore their own origins, both Indo-European and later Germanic.
>
> The problem was that there were two "schools" to choose from in exploring native European origins. First there was the standard academic version, which, as I say, was certainly de-emphasized by mainstream archaeology and things would remain this way until the late 1960s. The other alternatives were the occult ideas of Madame H. P. Blavatsky, as reinterpreted by Guido von List and Lanz von Lebenfels, whose ideas found a home not in German science but in the Ahnenerbe, a Nazi organization which specifically sought Germanic-Aryan origins.
>
> The Ahnenerbe became part of the SS in about 1940 and sent out expeditions of archaeologists and explorers in an attempt to establish a cultural and racial origin. Among these were the expeditions made to Tibet. Perhaps the Ahnenerbe thought the origins of the Germanics or the Aryans might be in Tibet, and so the grounds for measuring the populace in terms of typological anthropology, but this was not the view of the top scientists in German anthropology at the time, like Hans F.K. Guenther or Egon Freiherr von Eickstedt. So

> please, let us not ever confuse the Ahnenerbe with German science other than the fact that the scientists employed by the Ahnenerbe were willing to be used for this purpose. This is certainly not unique to the Ahnenerbe—American anthropology and indeed all social science was and still is chock-full of such prostitutes who are directly or indirectly funded by the government through academia and dare not deviate from current political correctness [lest they] lose their jobs.[46]

But even if the Ahnenerbe entertained this racial-origin hypothesis on their first trip to Tibet, after that trip the reasons for continued expeditions must have been quite different. The Tibetans had asked for weapons and the Germans were probably happy to give them weapons. In exchange the Germans would get a secret airbase somewhere in Tibet.

Tibetan monks did make their way to Germany somehow and it was probably by aircraft. A letter was brought to Germany from the high Lama, and it seemed that Tibet had a pro-German stance at the outbreak of World War Two.

Stevens then theorizes that one of the things that the SS was interested in were the stories of the flying vehicles called vimanas in various Indian and Tibetan texts.[60] Stevens refers to the German writer Dr. Axel Stoll (*Hoch-Technologie im Dritten Reich*, 2001) who writes that the SS was interested in the various vimana texts:

> Dr. Axel Stoll makes the point that the origins of some of the German flying disc technology may have been ancient India. Nobody really denies this is a possibility. Dr. Stoll, it seems, has really taken an interest in this possibility, translating the Vymaanika Shastra (evidently from an American/English version) into a scientific understanding in the German language. The point for us is that Dr. Stoll believes that it may be true that this knowledge survives and is guarded somewhere to this day, and suggests this place may be Tibet.
>
> At a later point Dr. Stoll returns to Tibet in the book referenced above and describes it as a possible place for a

> secret German base. Here, he says outright that the Tibetans fought for Germany during the Second World War, so a good connection and rationale for such a base, at least in his mind, is evident. He says this would be an underground base and it would have been secured using electromagnetic means. This is what I have called the "electromagnetic vampire" in my book, *Hitler's Suppressed and Still-Secret Weapons, Science and Technology*. In chapter five of that book is a description of Dr. Stoll's electromagnetic security system. Suffice it to say that Dr. Stoll has his own unified field theory and in his interpretation, life force can simply be sucked out of an intruder using a sort of reverse-Faraday cage and the right electronics. In this underground secured base, among all the other things sequestered there, Dr. Stoll envisions flying discs.
>
> So, the Nazis were interested in Tibet, possibly as a storehouse of ancient knowledge awaiting re-discovery through the efforts of the Ahnenerbe. We have evidence in the monks that people in some numbers were brought out of Tibet. It would have been almost impossible to smuggle them through India as Dr. Farrell points out. Dr. Stoll believes German underground base(s) would have been built, and that they would have contained, among other things, flying discs.[46]

So, now we not only have long-range aircraft based in Tibet but flying saucers as well. Stevens tells us that German flying discs, even the conventional ones, are credited by all German sources as being able to handle the Himalayas. He says the engineer Rudolf Lusar, who worked at the German Patent Office, gives a German flying disc the ability to climb, in three minutes, to an altitude of 12,400 meters or over 40,000 feet. Such a flying disk could hop over the Himalayas and Mt. Everest with ease. Flying from Tibet to anywhere in India after the war would not be a problem, since in the immediate postwar years the British had no real air defenses set up in India. Stevens thinks a secret base in Tibet would have housed flying saucers that were used to travel around Asia, as they were not really weapons. This secret Tibetan base may also be the source of

A map of Tibet showing three possible sites of a Nazi airbase with a Black Sun.

the curious photos of flying saucers taken in China during WWII.

A Strange UFO Incident in Karachi

That this secret SS airbase was still active in Tibet in 1955 can be surmised by the curious UFO incident in Karachi, Pakistan, on March 23, 1955. The story of an obviously man-made flying disc comes from the late 90s British publication *UFO Magazine* in an article that is entitled "Incident at Karachi."

Pakistan had separated from India several years after the end of WWII in August of 1947 when the two countries gained self-rule from Great Britain. Karachi is located on the plains of Pakistan, to the southwest of Tibet. On that day in 1955 thousands of people in Karachi saw a huge flying disk near the airport. The disc was first observed by British military pilots and men who were returning from Singapore to Britain via Bangkok and Karachi. Many of these men were at 3,000 feet, aboard a DC-9 attempting to land at Karachi airport. Says Stevens about the report in *UFO Magazine*:

> The teller of this tale is Mr. Frank J. Parker, who in the late 1990s was living in Hyde, Cheshire, England. Mr. Parker, twenty years old at the time, was on board the DC-9 in question and so first observed the flying saucer after some excited talk in the passenger cabin. Because the passengers from one side of the aircraft were standing up and moving to the other side of the aircraft, the captain ordered all the

> passengers back to their seats to avoid imbalance of the aircraft. Well, I have flown on a DC-6. It was a large, four-engine passenger aircraft. If it was in danger of becoming imbalanced, there were certainly many individuals moving from one side to the other observing this flying craft, and so it must have caused great excitement.
>
> Upon landing, the military men boarded a bus for the 16- or 17-mile drive to Karachi. They could still see the flying saucer, stationary and suspended over the city. At first, according to Parker, from the air it looked like an orange ball, but took on a disc shape as they drew nearer to it. As they continued on toward it, Parker estimated that it was 300-400 feet in the air. The bus turned a corner on to "the parade grounds" where Parker describes the sight that met his eyes as "incredible." Thousands of people, mostly Pakistanis, were amassed on the parade grounds, and above them, as if suspended, was the flying disc. Many of the Pakistanis knelt on the ground, hands clasped as though they were praying. Parker and some companions exited the bus and began moving through the crowd towards the spot over which the flying disc hung. Nobody was afraid or upset, according to Parker, and so the British party was able to move through the kneeling crowd without taking their eyes off the flying disc. They came to a halt directly below the still-suspended disc, now only about 60 yards above them.[46]

It is interesting to note here that the Pakistanis, all Muslims, were kneeling on the ground and praying. For them it was a sign from Allah. How many other historic incidents have been created by a brightly lit circular craft coming down from the sky?

Frank Parker then describes the scene before them:

> This had to have been at least 150 feet across and two-to-three stories high—by that I mean about 75 feet, it was incredible. Because it was hovering about 12-feet off the ground, we could see every detail. This was a 'nuts and bolts' spacecraft that looked man-made, because we could see what I could only describe as rivets, only perfectly aligned.

> I suppose because of its shape you could say it was a traditional 'flying saucer,' like two bronze cymbals, one on top of the other. It was magnificent, made up entirely of this metallic bronze-colored metal that applied to the catwalks, ladders, portholes and aerials. These were all fixed to the exterior of the craft. At the underside, we saw a big black hole, which interested me because whatever was holding this thing up had to come from there. Well, we saw what looked like fans and I counted eight rotors, but every few seconds there would be a 'whooshing' sound and that created sparks and an array of laser-like effects.

With catwalks, ladders, portholes, aerials, and rotors generating a downdraft, we can pretty much say that this craft was not some interstellar spacecraft. Parker continues, discussing the "whooshing" sound:

> Whenever this occurred, the entire craft would swing to the left and right and then come back again. This happened so many times that I believed whoever was behind the controls was experiencing engine problems... Every now and again, the craft ejected smoke towards the ground, and this gave off a strong smell of oil, but here's another curious thing—this stuff actually never reached the ground—rather it just seemed to evaporate... The craft made a constant noise not dissimilar to jet-engine propulsion, but nothing like as loud. This and the noise of the rotors would sometimes sound rough, as though they were faltering occasionally. Anyway, every now and again, it would force hot air downwards and this had the effect of blowing dust and pieces of paper all over the place.[46]

This fascinating incident would appear to involve one of the German flying disks coming from Tibet. What it was doing in the area is difficult to determine. It would seem to have been on a mission to simply harass the British forces in Karachi, who mainly serviced transport aircraft carrying soldiers back to Britain from India and the Far East. Perhaps it was there to pick someone up or let people off in a park where they could just melt into the citizenry of the city. From Karachi one could easily get trains, ships and airplanes to other parts of the world. Once again, western Tibet would be a good geographical spot for this secret airbase to exist, and the Chang Tang Highlands of the north section of western Tibet is an area that is extremely remote, even today, with few roads or towns. It should be noted here that except for water, the Germans would have needed very little from the surrounding terrain. The more remote, the better (as we see with the Arctic bases and Antarctica). Like these bases, the base in Tibet would have been supplied from the air, with aircraft—and incredibly, flying saucers—bringing machinery, operators, parts and food to the secret location.

Indeed, it is often asked, "If the Germans had invented flying disks at the end of the war, why didn't they win the war?" The simple answer to this is that they were only deployed after the war and were not really weapons, though some of these craft did have a cannon that was similar to a tank cannon. In a way they were flying tanks, but having them would not have won the war, with Germany being overwhelmed by armies from the east and west. These craft could fly very fast and could give ordinary aircraft quite a display of aerial

A photograph of a flying saucer taken in 1942 in Tientsin, China. Tientsin is now called Tianjin and is the main port for Beijing. Is it from the secret base in Tibet?

zip. World War II did not see vertical take off and landing (VTOL) aircraft until the final weeks of the war against the Japanese, in Asia when the first helicopters were used in northern Burma to rescue downed airmen.

Flying disks are far superior to helicopters. Both can be similarly armed, but the advantage flying disks have of being super quick transport and having the ability to land in virtually any open space that is large enough to accommodate the disk. This could be a city park, a rural farm, the top of a mountain or mesa, or even an empty highway or parking lot. When we look at the many early UFO reports from the 40s, 50s and 60s we see flying saucers doing exactly this. Landing in dark fields and disgorging passengers who walk to waiting black sedans. In many of these cases, Men in Black are waiting for them and whisk them away to destinations unknown.

One final note on the Nazi base in Tibet is that the final two episodes of the popular 1960s television show *The Man from UNCLE* featured a plot about a secret base in the Himalayas (Tibet). It is curious that the final episodes of the show, *The Seven Wonders of the World Affair* (parts one and two), were about a hidden base being operated by a secret organization named THRUSH, a thinly veiled Odessa-style group of ex-Nazis and criminals who were in Tibet for the purpose of attempting to control the world. In *The Man from UNCLE* series, agents from both Russia and the United States attempt to keep world order and the bad ex-Nazis from creating problems around the world. That the final episodes of the program, shown in January of 1968, were about a secret base in Tibet would seem to be a nod to the belief within intelligence agencies that the Germans were using a hidden base on the Tibetan plateau up until 1955 and possibly later.

Silver Balls Floating in Air Nazis' Newest War Device

1944

(The Associated Press)

Paris, Dec. 13.—As the Allied armies ground out new gains on the western front today, the Germans were disclosed to have thrown a new "device" into the war—mysterious silvery balls which float in the air.

Pilots report seeing these objects, both individually and in clusters, during forays over the Reich.

(The purpose of the floaters was not immediately evident. It is possible that they represent a new anti-aircraft defense instrument or weapon.)

(This dispatch was heavily censored at supreme headquarters.)

Secret Weapon Resembles Yule Decoration

PARIS (INS)—The Germans on the western front have produced a "secret" weapon in keeping with the Christmas season, it was disclosed officially Wednesday.

The new device, apparently an air defense weapon, resembles the huge glass balls which adorn Christmas trees.

They hang in the air sometimes singly, sometimes in clusters. They are colored silver and other shades and are apparently transparent.

No information was available as to what holds them up like stars in the sky, what is in them, or to their purpose.

Lately, they have been seen several times floating over German territory.

Floating Mystery Ball Is New Nazi Air Weapon

SUPREME HEADQUARTERS, Allied Expeditionary Force, Dec. 13—A new German weapon has made its appearance on the western air front, it was disclosed today.

Airmen of the American Air Force report that they are encountering silver colored spheres in the air over German territory. The spheres are encountered either singly or in clusters. Sometimes they are semi-translucent.

SUPREME HEADQUARTERS, Dec. 13 (Reuter)—The Germans have produced a "secret" weapon in keeping with the Christmas season.

The new device, apparently an air defense weapon, resembles the huge glass balls that adorn Christmas trees.

There was no information available as to what holds them up like stars in the sky, what is in them, or what their purpose is supposed to be.

In December of 1944 the Allies began to encounter mysterious balls floating in the air around their bombers and escorts over Germany. The Americans began to call them Foo Fighters, a name taken from the Smokey Stover comic strip. They may have encountered flying saucers as well.

Chapter Seven

Secret Cities in South America

If you want to find the secrets of the universe,
think in terms of energy, frequency and vibration.
—Nikola Tesla

Squadron 200 and the Secret Flights from Norway

Let us look now at Squadron 200 and the secret flights from Norway to Japan and from Europe to the Canary Islands. Stevens says that that the Ju-390 flew to Japan, over the polar route, from a secret base in Norway. The Ju-390 left Norway from a base in the far north at Bardufoss, flew over the north pole, over the Bering Strait, then down the east side of the Kamchatka Peninsula to the island of Pamushiro which was within the Japanese Empire. From there it flew over Japanese-controlled Manchoutikuo to Tokyo. Stevens says confirmation of this flight comes from a radio report by the Japanese Attachés for Marine Aircraft in Germany, dated March 21, 1945. Says Stevens:

> The purpose was to transport German high-tech weaponry secrets and some personnel involved with this work. This was exactly the same purpose as the famous U-234 voyage and the two methods of communication can be considered complimentary, each part of a larger whole. There may have been and probably were more flights involving the Ju-390 to Japan just as there were probably other U-boats delivering technology to Japan besides U-234.

Stevens says that the Focke-Wulf Fw-200 Condor was also used in the many long-range flights. This was originally a four-engine passenger aircraft but the Luftwaffe adapted it for long-range reconnaissance and even as an anti-shipping bomber. Besides extensive work in the North Sea and polar regions surrounding Norway, this aircraft also transported supplies all the way to Stalingrad in 1942.

Stevens also says that in 1944 there was a conference at Strasbourg of SS officers and other Nazi officers at which it was agreed that blueprints, machine tools, secret weapons, specialty steel, gold, money and scientists were to be transported outside the Reich for future use. The means of this transfer was the deployment of Geschwader 200 (Squadron 200), an elite flying group of young men, many of whom had lost their families. British intelligence was known to use orphans as part of their most dangerous operations (Agent 007 is said to be an orphan by author Ian Fleming). On both sides, the most dangerous missions were best done by those without families.

Geschwader 200 was the first choice of the Luftwaffe for truly dangerous missions. Geschwader 200 was, at one point, scheduled to fly the manned version of the V-1 rocket to high value targets. This would have certainly meant the loss of almost all of the pilots, as the men themselves knew. They were, briefly, the Kamikaze squadron of the Third Reich. This project was abandoned but Geschwader 200 was involved in a suicide attack on a bridge spanning the Oder River in the final stages of the war in an attempt to slow the Soviet advance. Says Stevens about the transfer of people and goods out of the Third Reich:

> In the transfer specified in the Strasbourg agreement, the goods and people necessary were loaded aboard Condor aircraft by Geschwader 200 and flown from points within the Reich to Madrid, then on to Cadiz. Cadiz is on the Mediterranean Sea. From there transport to South America was accomplished using both Fw-200 Condors of Geschwader 200 and U-boats. Additionally, the Azores or the Canary Islands were used as a stopover base.

These Focke-Wulf Condors could make long journeys, but could not fly as far as the Amerikabomber-type planes, such as the Ju-390, being developed at the end of the war. But flights to Spain and then to the Canary Islands or the Spanish Sahara were easily within their range, and the Condor could even fly from the Canary Islands to South America where they would probably land at a private airfield in Argentine Patagonia where Germans had large tracts of land.

Finally, Stevens says that the researcher/writer Friedrich Georg claimed the following tasks for the Ju-390 aside from being a long-range bomber:

1. Secret intelligence operations.
2. An escape vehicle for the Nazi leadership.
3. A long-distance delivery vehicle connecting to Japan.

Stevens mentions that the idea of giving the design for the long-range plane to the Japanese is included in number three above. The Nazis were happy to give the Japanese their technology and the Japanese would have found the Ju-390 aircraft useful, as it would have traversed the Pacific just as it would the Atlantic. Stevens also mentions that the SS commander Dr. Hans Kammler was said to have escaped Germany at the end of the war in a Ju-390, supposedly flying to South America. Alternatively, Kammler is said to have committed suicide, among various other accounts of his demise, none of which have ever been proven. Kammler is literally said to have died in a dozen different ways. Did he in fact escape to South America? Kammler was in charge of special projects for the SS and would have been quite familiar with the Ju-390. Himmler is also said to have escaped from Germany in a captured British plane flown out of a special SS airfield near Berlin. He allegedly flew to Spain and then made the journey to South America, probably by U-boat. In this version of his fate, a double of Himmler killed himself with a cyanide pill when he was captured by British troops at the end of the war, as the official story goes.

The Vril Society and Vril Craft

The Vril Society is said to have emerged from the Thule Society. The existence of a Vril Society was first alleged in 1960 by Jacques Bergier and Louis Pauwels in their book *The Morning of the Magicians*.[29] In the book they claimed that the Vril Society was a secret community of occultists in pre-Nazi Berlin that was a sort of inner circle of the Thule Society, of which Rudolf Hess was a member. They also thought that it was in close contact with the English group known as the Hermetic Order of the Golden Dawn.

According to such authors as Jocelyn Godwin[35] and Nicholas Goodrick-Clarke,[28] the Thule Society was originally a "German study group" headed by Walter Nauhaus, a wounded World War I veteran turned art student from Berlin who had become a keeper of pedigrees for the Germanenorden (or "Order of Teutons"), a secret society founded in 1911 and formally named in the following year.

Secret societies were booming in Germany during this period and Nauhaus moved to Munich in 1917 where his "Thule Society" was to be a cover name for the Munich branch of the Germanenorden, but a schism in the Order caused events to develop differently. In 1918, Nauhaus was contacted in Munich by Rudolf von Sebottendorf who was an occultist and newly elected head of the Bavarian province of the schismatic offshoot known as the Germanenorden Walvater of the Holy Grail. The two men became associates in a recruitment campaign, and Sebottendorff adopted Nauhaus's Thule Society as a cover name for his Munich lodge of the Germanenorden Walvater at its formal dedication on August 18, 1918.

"Thule" was a land located by Greco-Roman geographers in the farthest north (often displayed as Iceland). The Latin term "Ultima Thule" is mentioned by the Roman poet Virgil in his pastoral poems called the *Georgics*. Thule may have been the original name for Scandinavia, although Virgil simply uses it as an expression for the edge of the known world to the north. The Thule Society identified Ultima Thule as a lost ancient landmass in the extreme north, near Greenland or Iceland, said by Nazi mystics to be the capital of ancient Hyperborea. Atlantis fitted into these myths as well and the Nazi mystics believed that Atlantis had existed in the Northern Sea, rather than in the Azores or the Caribbean.

Wikipedia says that Hitler biographer Ian Kershaw said that the organization's "membership list ...reads like a Who's Who of early Nazi sympathizers and leading figures in Munich," including Rudolf Hess, Alfred Rosenberg, Hans Frank, Julius Lehmann,

Gottfried Feder, Dietrich Eckart, and Karl Harrer. Wikipedia says that in his book *Monsieur Gurdjieff*, Louis Pauwels claims that a Vril Society was "founded by General Karl Haushofer, a student of Russian magician and metaphysician Georges Gurdjieff."

Author Nicholas Goodrick-Clarke[28] contends that Rudolf Hess and Hans Frank (a German politician and lawyer who served as head of the General Government in Poland during the Second World War and was executed at Nuremburg in October 1946), had been Thule members, but other leading Nazis had only been invited to speak at Thule meetings. It is generally thought that the Thule Society itself dissolved before 1930, but the Vril Society continued on until 1945.

Vril itself was a fabled source of free and infinite energy, a sort of electricity all around us that could be utilized, similar to static electricity. The term vril was first used in the 1871 novel by Edward Bulwer-Lytton, *Vril: The Coming Race*.[53] The book is about an inner-earth dwelling race of superhumans called the Vril-ya, who master vril energy for both its healing and destructive properties. The book warns that one day the Vril-ya superhumans will rise to the surface of the Earth and destroy the inferior race of *Homo sapiens*.

The book and the concept of vril inspired the creation of the Vril Society out of the Thule Society. It was taught that vril was a Cosmic Primal Force—an exotic spiritual technology that would bring about a new Utopian era for humanity: the Rebirth of Atlantis.

These themes appeared in two booklets published in Berlin in 1930 by an obscure group called Reichsarbeitsgemeinschaft Das Kommende Deutschland (Imperial Working Society the Coming Germany), or RAG for short. RAG taught its readers to meditate on the image of an apple sliced vertically in half, representing the map of universal free energy available from Earth's magnetic field. Using this knowledge they would harness vril, the "all-force of the forces of nature" by using "ball-shaped power generators" to channel the "constant flow of free radiant energy between outer space and Earth."

The booklets describe Atlantean dynamo-technology superior to modern mechanistic science, saying that their spiritual technology and limitless vril energy is what enabled Egyptians and Mayans to build massive pyramids.

One of the more popular legends is of the Vril Maidens—an inner circle of young female psychic mediums, led by a beautiful woman named Maria Orsic. These women kept their hair long because they believed it to be an extension of the nervous system and acted as an antenna when telepathically communicating between worlds.

Nicholas Goodrick-Clarke tells us in his scholarly book *Black Sun: Aryan Cults, Esoteric Nazism, and the Politics of Identity*,[28] that the story began in 1919 when an inner group of Thule and Vril Society members held a meeting in Vienna, where they met with a psychic named Maria Orsic. She allegedly presented transcripts of automatic writing she received in a language that she didn't understand. The language was found to be ancient Sumerian, allegedly channeled from a planet in the Aldebaran solar system—Aldebaran being the brightest star in the constellation of Taurus—68 light years away from Earth.

The Vril Maidens channeled blueprints for time travel machines, anti-gravity technology, and more. Over the next decade this led to the development of the Vril and Haunebu series of anti-gravity flying saucers.[28]

According to information on a "vril site" that promotes Maria Orsic on the Internet, the Vril Society was formed by a group of female psychic mediums led by the Thule Gesellschaft medium Maria Orsitsch (Orsic) of Zagreb. She claimed to have received communication from Aryan aliens living on Alpha Centauri, in the Aldebaran system. Allegedly, these aliens had visited Earth and settled in Sumeria, and the word vril was formed from the ancient Sumerian word "Vri-Il" ("like god"). Says the text:

> A second medium was known only as Sigrun, a name etymologically related to Sigrune, a Valkyrie and one of Wotan's nine daughters in Norse legend. The Society allegedly taught concentration exercises designed to awaken the forces of Vril, and their main goal was to achieve Raumflug (Spaceflight) to reach Aldebaran. To achieve this, the Vril Society joined the Thule Gesellschaft to fund an ambitious program involving an inter-dimensional flight machine based on psychic revelations from the Aldebaran aliens. Members of the Vril Society are said to have included Adolf Hitler, Alfred Rosenberg, Heinrich

An alleged photo of the Vril priestess called Maria Orsic.

> Himmler, Hermann Göring, and Hitler's personal physician, Dr. Theodor Morell. These were original members of the Thule Society which supposedly joined Vril in 1919. The NSDAP (NationalSozialistische Deutsche ArbeiterPartei) was created by Thule in 1920, one year later. Dr. Krohn, who helped to create the Nazi flag, was also a Thulist. With Hitler in power in 1933, both Thule and Vril Gesellschafts allegedly received official state backing for continued disc development programs aimed at both spaceflight and possibly a war machine.
>
> After 1941 Hitler forbade secret societies, so both Thule and Vril were documented under the SS E-IV unit. The claim of an ability to travel in some inter-dimensional mode is similar to Vril claims of channeled flight with the Jenseitsflugmaschine (Other World Flight Machine) and the Vril Flugscheiben (Flight Discs).

It is difficult to verify this information on the Vril Society or Maria Orsic. Some believe that she was born in Zagreb on October 31, 1894 and visited Rudolf Hess in Munich in 1924. She is said to have disappeared in 1945. An Internet biography of her states:

> In December 1943 Maria attended, together with Sigrun, [another psychic, even more mysterious than Orsic] a meeting held by Vril at the seaside resort of Kolberg. The main purpose of the meeting was to deal with the Aldebaran project. The Vril mediums had received precise information regarding the habitable planets around the sun Aldebaran and they were willing to plan a trip there. This project was discussed again the 22nd January 1944 in a meeting between Hitler, Himmler, Dr. W. Schumann (scientist and professor in the Technical University of Munich) and Kunkel of the Vril Gesellschaft. It was decided that a *Vril 7 Jaeger* would be sent through a dimension channel independent of the speed of light to Aldebaran. According to N. Ratthofer (writer), a first test flight in the dimension channel took place in late 1944.
>
> Maria Orsic disappeared in 1945. The 11th of March

> of 1945 an internal document of the Vril Gesellschaft was sent to all its members; a letter written by Maria Orsic. The letter ends: “niemand bleibt hier” [no one stays here]. This was the last announcement from Vril. It is speculated they departed to Aldebaran.

It seems that the Vril Society was real but the stories about Maria Orsic and the contact with the planet Aldebaran are doubted by many including Nicholas Goodrick-Clarke.[28] Some researchers claim that nothing can be found about Maria Orsic before 1990. However, it does seem that one of the saucer craft built by the Germans was given the name Vril. Henry Stevens says:

> The late Heiner Gehring wrote to me that he had been told that the efforts of the Vril Society at channeling continued after the war, until 1946. It is no secret that in the lore propounded by Ralf Ettl and Norbert Juergen-Ratthofer elements of the Nazi regime allegedly associated with the Vril Society launched an interdimensional flight to an extraterrestrial world seeking help in the Second World War. This flight communicated, allegedly, via medial transmission (or channeling). This medial contact, according to Heiner’s source, continued until 1946 with the living crew of that ship when it was abruptly terminated. Of course, there is no way to verify any of this.[46]

The Vril Disks

Since about 1990, researchers in Europe, Australia, America and elsewhere have received documents that show the designs of German flying saucers. A number of curious photos of the craft, in flight and on the ground, have also surfaced.

Henry Stevens says that the flying craft type called “Haunebu” is a postwar creation, at least in its operational version. He says a Haunebu craft is a large-size flying saucer, at least 30 feet or so in diameter, with a classic dome on its upper side, and without any indication of rotating discs, wheels or parts.

There are a number of diagrams and photos of these craft. Stevens says that the alleged photographs of Vril craft and Haunebu

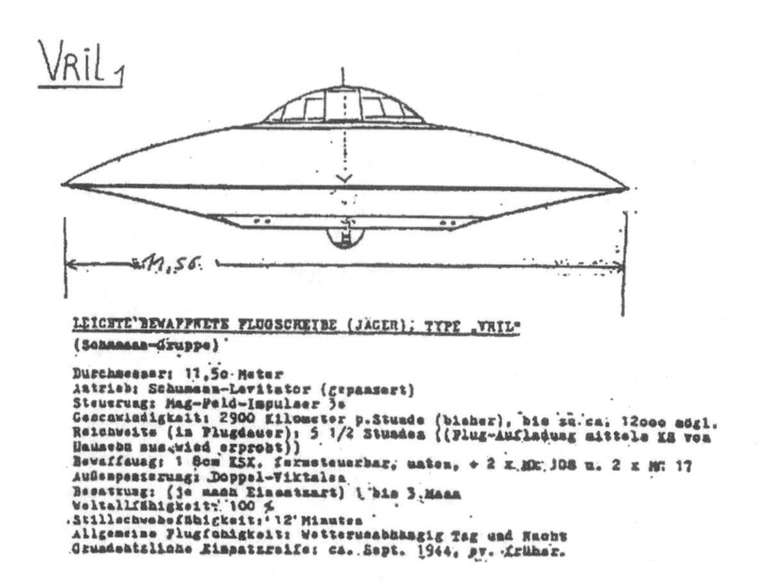

LEICHTE BEWAFFNETE FLUGSCHEIBE (JÄGER); TYPE „VRIL“
(Schumann-Gruppe)

Durchmesser: 11,5o Meter
Antrieb: Schumann-Levitator (gepanzert)
Steuerung: Mag-Feld-Impulser 3e
Geschwindigkeit: 2900 Kilometer p.Stunde (bisher), bis zu ca. 12000 mögl.
Reichweite (in Flugdauer): 5 1/2 Stunden ((Flug-Aufladung mittels KS von Hauneburg aus wied erprobt))
Bewaffnung: 1 8cm KSK, fernsteuerbar, unten, + 2 x MK 108 u. 2 x MG 17
Außenpanzerung: Doppel-Viktalen
Besatzung: (je nach Einsatzart) 1 bis 3 Mann
Weltallfähigkeit: 100 %
Stillschwebefähigkeit: 12 Minuten
Allgemeine Flugfähigkeit: Wetterunabhängig Tag und Nacht
Grundsätzliche Einsatzreife: ca. Sept. 1944, ev. früher.

VRIL 2 und VRIL 9 : Projektiert, aber nicht gebaut

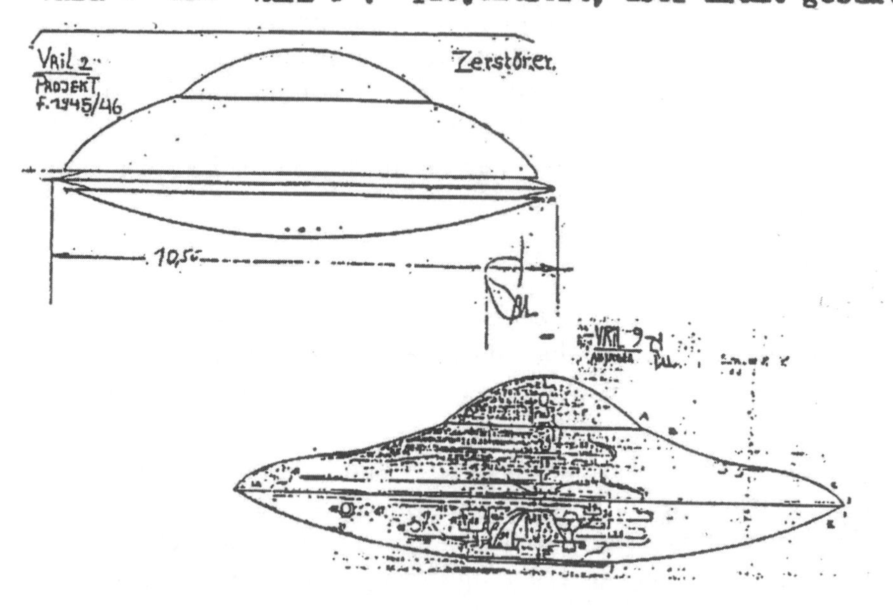

One of the documents containing plans for the Vril craft.

craft, that are said to date from the war years, are all from one source. This one source was an anonymous package dropped off in England for researcher Ralf Ettl, who was working there on an unrelated project at the time. This package of photos arrived at Ettl's abode in the late 1980s and he then distributed the photos to various other researchers in Europe and around the world. Ettl has written and co-written a number of books in German including a book titled *Das Vril-Projekt*. These documents are thought to be genuine and came from secret SS files kept in Austria.

The files and photos describe two types of flying saucer, the Vril, a small two-man craft and the Haunebu, a larger saucer-shaped craft that had seats for nine people. A third type of craft is shown, the classic cigar-shaped mothership called the Andromeda. This craft could allegedly hold one Haunebu II craft and four of the smaller Vril craft.

The projects were supposedly under the supervision of the "Vril-Gesellschaft" and of the SS E IV (a secret development center for alternative energy of the SS), and as such they were not directly under Hitler's and the Nazi Party's orders and were not really planned for war use. But later when Germany's situation deteriorated the SS began to think about using the flying disks in the war. They were originally scout craft. Later they were used for other purposes.

Supposedly, the Vril and Haunebu craft had these statistics:

Vril 1 from Sept. 44
Diameter 11.5 m
Drive: Schuman levitator (antigravitation eqpm.)
Steuerung/steering: mag-field-impulser
Velocity: 2900-12000km/h
Capacity: 5.5 h in air

Haunebu I from Dec. 44
Diameter 25 m
Drive: Thule tachyonator 7b (antigravitation eqpm.)
Steuerung/steering: mag-field-impulser
Velocity: 4800-17000km/h
Capacity: 18 hrs in air
Crew 8 people

Haunebu II from 43-44
Diameter 26.3 m
Drive: Thule tachyonator 7b (antigravitation eqpm.)
Steuerung/steering: mag-field-impulser
Velocity: 6000-21000km/h
Capacity: 55 h in air
Crew 9 people

Haunebu III from sometime in 45
Diameter 71 m
Drive: Thule tachyonator 7b and Schuman levitators (antigravitation eqpm.)
Steuerung/steering: mag-field-impulser
Velocity: 7000-40000km/h
Capacity: 8 weeks in air
Crew 32 people

According to the Vril Society websites on the Internet:

> In principle the "other side flying machine" should create an extremely strong field around itself extending somewhat into its surroundings which would render the space thus enclosed including the machine a microcosm absolutely independent of the earthbound space. At maximum strength this field would be independent of all surrounding universal forces—like gravitation, electromagnetism, radiation and matter of any kind—and could therefore manoeuver within the gravitational or any other field at will, without the

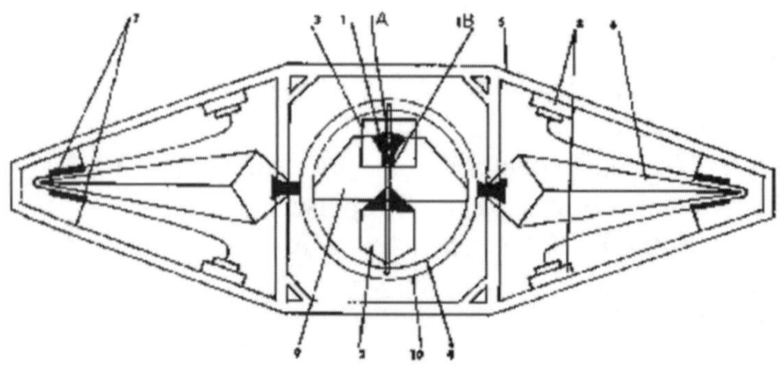

A diagram of the Vril craft.

Geheime Kommandosache!

Flugkreisel-Erprobung, Stand / Anzahl Erprobungsflüge:

HAUNEBU I (vorhanden 2 Stück)	52	E-IV
HAUNEBU II (vorhanden 7 Stück)	106	E-IV
HAUNEBU III (vorhanden 1 Stück)	19	E-IV
(VRIL I) (vorhanden 17 Stück)	84	(Schumann)

Empfehlung:
Beschleunigen von Abschlußerprobung
und Produktion „Haunebu II"
+ „VRIL I"

HAUNEBU I

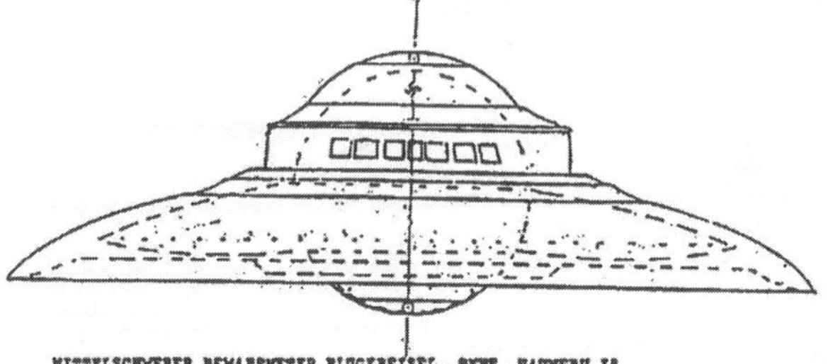

MITTELSCHWERER BEWAFFNETER FLUGKREISEL, TYPE „HAUNEBU I"

Durchmesser: 25 Meter
Antrieb: Thule-Tachyonator 7b
Steuerung: Mag-Feld-Impulser 4
Geschwindigkeit: 4800 Kilom.p.Std. (rechn. bis 17000)
Reichweite in Flugzeit: 18 Stunden
Bewaffnung: 2 x 8cm KSK in Drehtürmen und 4 x MK 108, starr nach vorn
Außenpanzerung: Doppel-Victalen
Besatzung: 8 Mann
Weltallfähigkeit: 60 %
Stillschwebefähigkeit: 8 Minuten
Allgemeine Flugfähigkeit: Tag wie Nacht
Grundsätzliche Einsatztauglichkeit: 60 %
Frontverfügbarkeit: Nicht vor Jahresende 44

Bemerkung: Die SS-E-IV hält Konzentration auf bereits im Versuch stehende „Haunebu II" für sinnvoller als an beiden Typen parallel weiterzuarbeiten. „Haunebu II" verspricht entscheidende Verbesserungen in nahezu allen Punkten. Höhere Herstellungskosten scheinen gerechtfertigt - besonders mit Blick auf Führer-Sonderbefehl, Flugkreisel betreffend.

Plans for the Haunebu I craft.

acceleration forces being effective or perceptible.

Before the end of 1934 the RFZ 2 was ready, with a Vril drive and a "magnetic field impulse steering unit." It had a diameter of five meters and the following flying characteristics: With rising speed the visible contours became blurred and the craft showed the colors typical for UFOs: depending on the drive setting red, orange, yellow, green, white, blue or purple. It worked—and it should meet a remarkable destiny in 1941, during the "Battle of Britain", when it was used as transatlantic reconnaissance craft, because for these flights the German standard fighters ME 109 had an insufficient range. By the end of 1941 it was photographed over the southern Atlantic on its way to the German cruiser *Atlantis* in Antarctic waters. It could not be used as a fighter though. The impulse steering allowed it only changes of direction at 90gr 45gr or 22.50 gr, but that is exactly the right-angled flying pattern associated with and typical for UFOs today!

After the success of the small RFZ 2 as a distant reconnoiter craft the Vril Gesellschaft got its own test area in Brandenburg. By the end of 1942 the lightly armed 'VRIL-1-Jager" (VRIL-1 fighter) was airborne. It measured 11.5 meters across, carried one person, had a "Schumann-Levitator" drive and a "magnetic field impulse steering unit." It reached speeds of 2,900 to 12,000 km/h, could change direction at a right angle at full speed without affecting the pilot, could fly in any weather and had a 100% space capability. Seventeen VRIL-1's were built and the same versions had two seats and glass domes.

Also during this time another project was worked on, the V-7. Several disks were built under this code, but with conventional jet engines. ANDREAS EPP had designed a combination of levitating disk and jet propulsion, the RFZ 7. The design groups SCHRIEVER-HABERMOHL and MIETHE-BELLUZO worked on it. The RFZ 7 had a diameter of forty-two meters; it crashed on landing at Spitzbergen. A second craft was later photographed outside Prague. According to Andreas Epp this craft was to

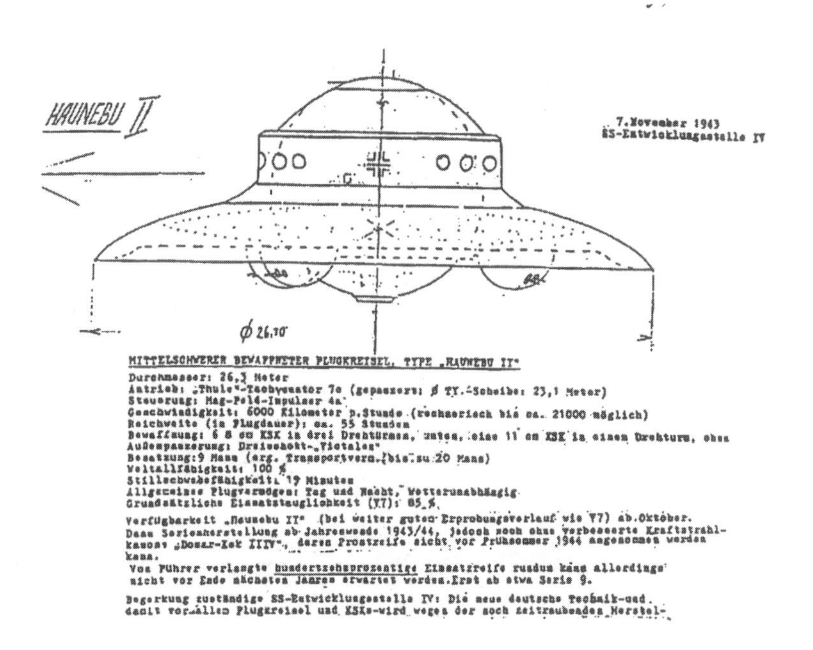

MITTELSCHWERER BEWAFFNETER FLUGKREISEL, TYPE „HAUNEBU II"

Durchmesser: 26,3 Meter
Antrieb: „Thule"-Tachyonator 7c (gepanzert: ø TY.-Scheibe: 23,1 Meter)
Steuerung: Mag-Feld-Impulser 4a
Geschwindigkeit: 6000 Kilometer p. Stunde (rechnerisch bis ca. 21000 möglich)
Reichweite (in Flugdauer): ca. 55 Stunden
Bewaffnung: 6 8 cm KSK in drei Drehtürmen, unten, eine 11 cm KSK in einem Drehturm, oben
Außenpanzerung: Dreischott-„Victalen"
Besatzung: 9 Mann (erg. Transportverm. bis zu 20 Mann)
Weltallfähigkeit: 100 %
Stillschwebefähigkeit: 19 Minuten
Allgemeines Flugvermögen: Tag und Nacht, wetterunabhängig
Grundsätzliche Einsatztauglichkeit (V7): 85 %

Verfügbarkeit „Haunebu II" (bei weiter gutem Erprobungsverlauf wie V7) ab Oktober. Dann Serienherstellung ab Jahreswende 1943/44, jedoch noch ohne verbesserte Kraftstrahlkanone „Donar-Kek XIIV", deren Frontreife nicht vor Frühsommer 1944 angenommen werden kann.

Vom Führer verlangte hundertzehnprozentige Einsatzreife rundum kann allerdings nicht vor Ende nächsten Jahres erwartet werden. Erst ab etwa Serie 9.

Bemerkung zuständige SS-Entwicklungsstelle IV: Die neue deutsche Technik und damit vor allem Flugkreisel und KSKs wird wegen der noch zeitraubenden Herstel-

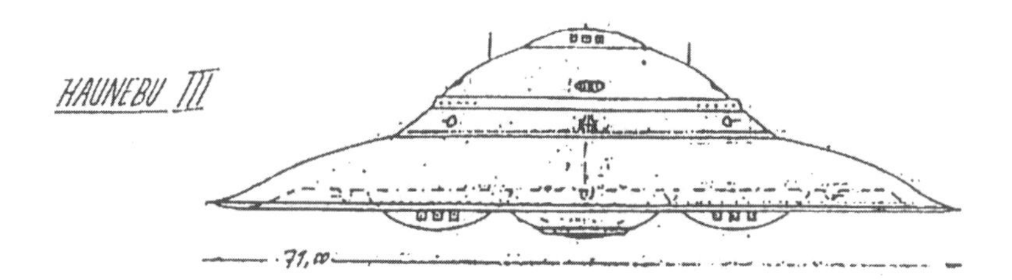

SCHWERER BEWAFFNETER FLUGKREISEL „HAUNEBU III"

Durchmesser: 71 Meter
Antrieb: Thule-Tachionator 7c plus Schumann-Levitatoren (gepanzert)
Steuerung: Mag-Feld-Impulser 4a
Geschwindigkeit: ca. 7000 Kilom. p. Stunde (rechnerisch bis zu 40000)
Reichweite (in Flugdauer): ca. 8 Wochen (bei E-L-Flug 40% mehr)
Bewaffnung: 4 x 11cm KSK in Drehtürmen (3 unten, 1 oben), 10 x 8cm KSK in Drehringen plus 6 x MK 108, 8 x 3cm KSK ferngesteuert
Außenpanzerung: Dreischott-Victalen
Besatzung: 32 Mann (erg. Transportverm. max. 70 Personen)
Weltallfähigkeit: 100 %.
Stillschwebefähigkeit: 73 Minuten.
Allgemeines Flugvermögen: Wetterunabhängig Tag und Nacht
Grundsätzliche Einsatztauglichkeit: Etwa 1945.

Bemerkung: SS-E-IV hält den Hinweis für notwendig, daß in „Haunebu III" ein großartiges Werk deutscher Technik im entstehen ist, wegen der allgemeinen Materiallage aber alle Kräfte auf das schneller verfügbare Haunebu II gesetzt werden sollten.
Gemeinsam mit dem leichten Flugkreisel „Vril" der Schumann-Gruppe könnte „Haunebu II" die vom Führer aufgestellten Forderungen sicherlich erfüllen.

Plans for the Haunebu II and III craft.

be armed with nuclear heads to attack New York. In July 1941 SCHRIEVER and HABERMOHL built a vertical takeoff round craft with jet propulsion, but it had severe shortcomings.

They went on to develop an "electro-gravitational flying gyro" with a "tachyon drive" which proved more successful. Then Schriever, Habermohl and Belluzo built the RFZ 7 T that was fully functional. The V-7 flying disks however were mere toys compared to the Vril and Haunebu disks. Within the SS there was a group studying alternative energy, the SS-E-IV (Development Group IV of the Black Sun) whose main task was to render Germany independent of foreign oil. The SS-E-IV developed from the existing Vril drives and the tachyon converter of Captain Hans Coler the "THULE DRIVE" which later was called the "THULE TACHYONATOR."

Indeed, a device called the "Coler Converter" does exist and British Intelligence released several documents on Hans Coler's

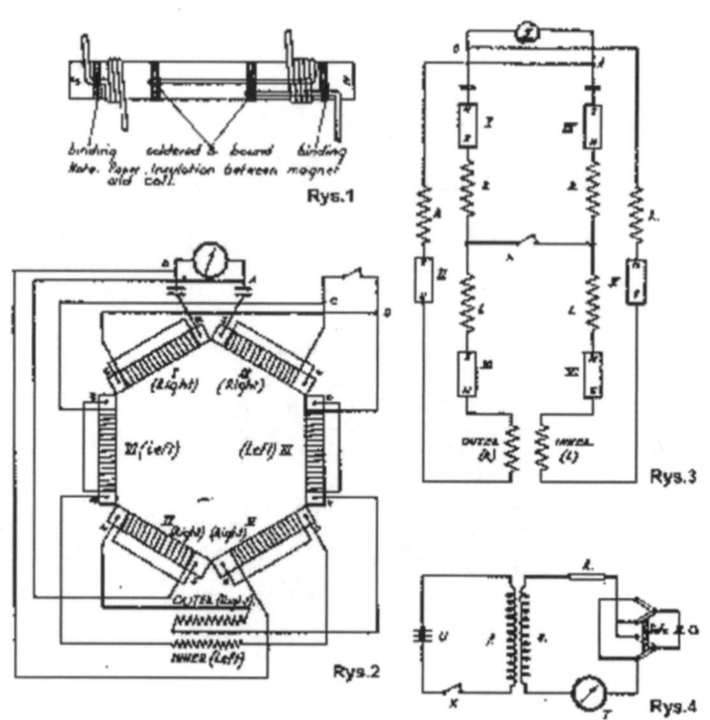

A diagram of the Coler Converter.

invention after the war. It is a device commonly discussed among free energy researchers and uses electromagnets in a special configuration as part of its function in creating power. The mention of the "electro-gravitational flying gyro" is very intriguing. It must be remembered that helicopters did not exist during most of WWII and a vertical take off and landing craft (VTOL) was very much desired during the war. The earliest experimental gyro copters and helicopters were manufactured and used in the middle of 1945. Importantly, as mentioned in the text above, Germany did not have oil resources and was very dependent on imported oil and other forms of imported energy. The exotic energy experiments that they funded were designed to make Germany independent of these foreign oil sources.

A Saucer Called Haunebu

The Haunebu craft are the most famous of the German saucers and there were allegedly three types built. Drawings and photos of the craft were published in various books in the 1950s, 60s and 70s, including the George Adamski books, but the more detailed diagrams and photos did not emerge until the document dump by

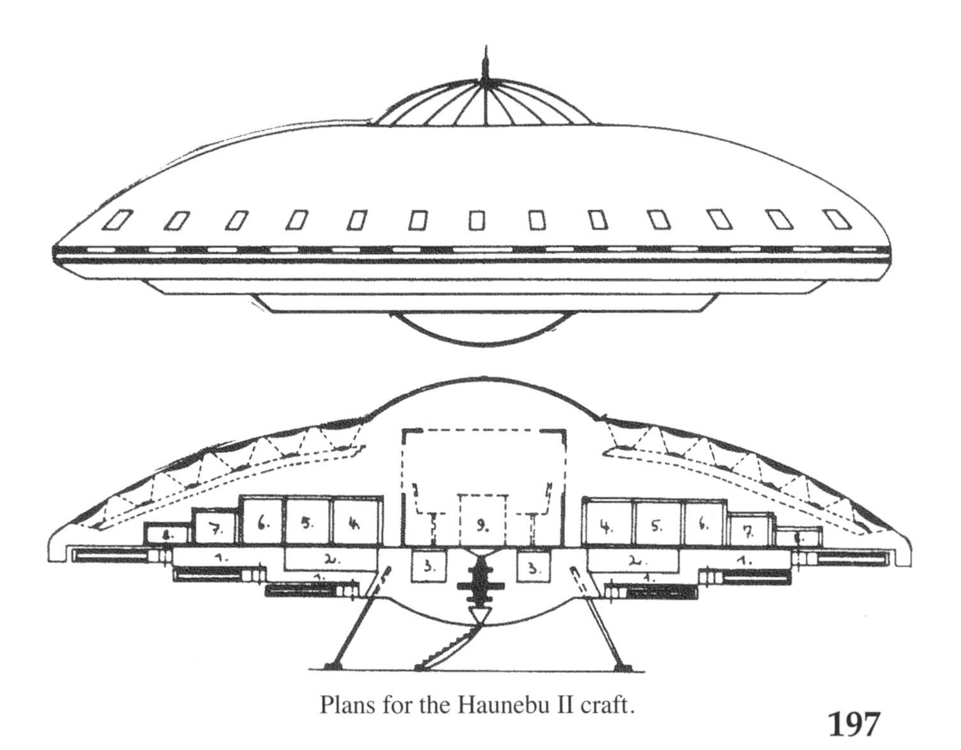

Plans for the Haunebu II craft.

Ralf Ettl in 1989. As in all clandestine programs, the names of the devices or craft were ultimately secret and the odd name Haunebu would have originally been a code word.

So, what is the meaning of the code word Haunebu? I found it very difficult to find out what this word meant. Unlike the term vril, no explanation of the word Haunebu is given in any of the literature that can be found on the subject. As with many code names, such as the British code name for their secret Antarctic excursion, Operation Tabarin, German code names did not necessarily have anything to do with the actual mission, device, or operation itself. This may be

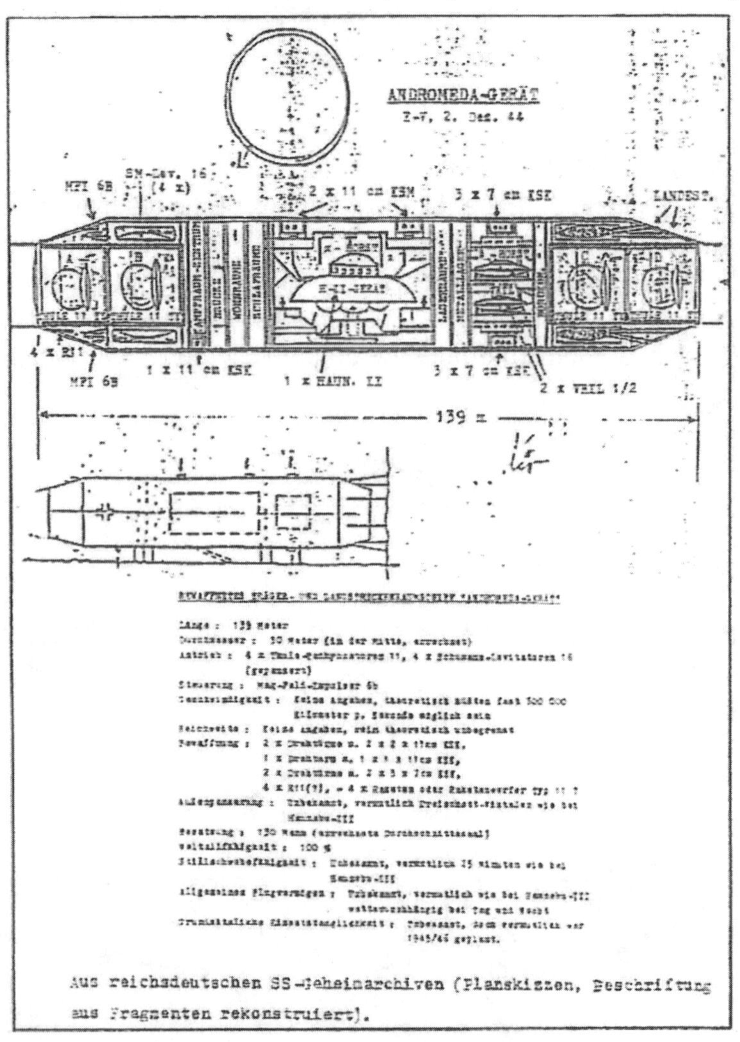

Plans for the Andromeda Craft, a tubular airship that held one Haunebu disk craft and several Vril disk craft.

the case with the word Haunebu—it may have some arcane meaning that only a few of the inner circle of the Vril Society would know. Or, the name might refer to a childhood dog as was the case with "Indiana" Jones—a famous moniker derived from that of the family pooch.

In looking it up on various dictionaries on the Internet, one lead seemed to show that it was a type of hair color. However, the best explanation that I was able to find was this from wiktionary.org:

> Haunebu (plural Haunebus or Haunebu)
> (*Egyptology*, plural "Haunebu") A member of a people from the Aegean Sea.
> (*ufology*, plural "Haunebus") Any of a class of flying saucers supposedly built by the Nazis.

So, the best explanation that I have been able to come up with for a meaning behind the code word "Haunebu" is that it refers to the original Greek raiders known to the Egyptians as the Sea Peoples who fought against Egypt circa 1200 BC and were ultimately repelled. These Sea Peoples were depicted as having horned helmets, like Vikings, and they overwhelmed coastal Egypt for a period until they were defeated. These people essentially settled in the Eastern Mediterranean in what is today Lebanon, Turkey and Greece. These people were, apparently, the "Haunebu." It has been suspected that the Sea Peoples were from Germany and Scandinavia and this could fit in with the SS's penchant for using esoteric and raider-type names such as vril, werewolf, death's head, storm trooper and so on. The SS Death's Head rings, such as the one Himmler wore, are now valuable collector's items and were worn by most of the U-boat officers and other commanders.

Supposedly, the Haunebu was an "armed flying gyro" that was first tested in 1939 with the Haunebu II being built in 1942-43, though many researchers think that Haunebu only came at the very end of the war.

A typical SS Death Head-Totenkopf ring.

According to the Vril websites on the Internet that promote the mysterious Maria Orsic and her contact with the planet Aldebaran, we have this curious timeline:

> In August 1939 the first RFZ 5 took off. It was an armed flying gyro with the odd name "HAUNEBU I." It was twenty-five meters across and carried a crew of eight. At first it reached a speed of 4,800 km/h, later up to 17,000 km/h. It was equipped with two 6 cm KSK ("Kraftstrahlkanonen," power ray guns in revolving towers) and four machine guns. It had a 60% space capability. By the end of 1942 the HAUNEBU II was ready. The diameters varied from twenty-six to thirty-two meters and their height from nine to eleven meters. They carried between nine and twenty people, had a Thule Tachyonator drive and near the ground reached a speed of 6,000 km/h. It could fly in space and had a range of fifty-five flying hours. At this time there existed already plans for a large-capacity craft, the VRIL 7 with a diameter of 120m. A short while later the HAUNEBU III, the showpiece of all disks, was ready, with seventy-one meters across. It was filmed flying. It could transport thirty-two men, could remain airborne for eight weeks and reached

A photo of a Haunebu II craft on the ground.

A photo of a Vril craft in flight.

at least 7,000 km/h (according to documents in the secret SS archives up to 40,000 km/h).

At the beginning of 1943 it was planned to build in the Zeppelin works a cigar-shaped mother ship. The ANDROMEDA DEVICE of a length of 139m should transport several saucer-shaped craft in its body for flights of long duration (interstellar flights).

By Christmas 1943 an important meeting of the VRIL-GESELLSCHAFT took place at the seaside resort of Kolberg. The two mediums Maria Ortic and Sigrun attended. The main item on the agenda was the ALDEBARAN PROJECT. The mediums had received precise information about the habitable planets around the sun Aldebaran and one began to plan a trip there. At a January 22, 1944 meeting between HITLER, HIMMLER, Kunkel (of the Vril Society) and Dr. Schumann this project was discussed. It was planned to send the VRIL 7 large-capacity craft through a dimension channel independent of the speed of light to Aldebaran. According to Ratthofer a first test flight in the dimension channel took place in the winter of 1944. It barely missed disaster, for photographs show the Vril 7 after the flight looking "as if it had been flying for a hundred years." The outer skin was looking aged and was damaged in several places.

On February 14, 1944, the supersonic helicopter—constructed by Schriever and Habermohl under the V 7 project—that was equipped with twelve turbo-units

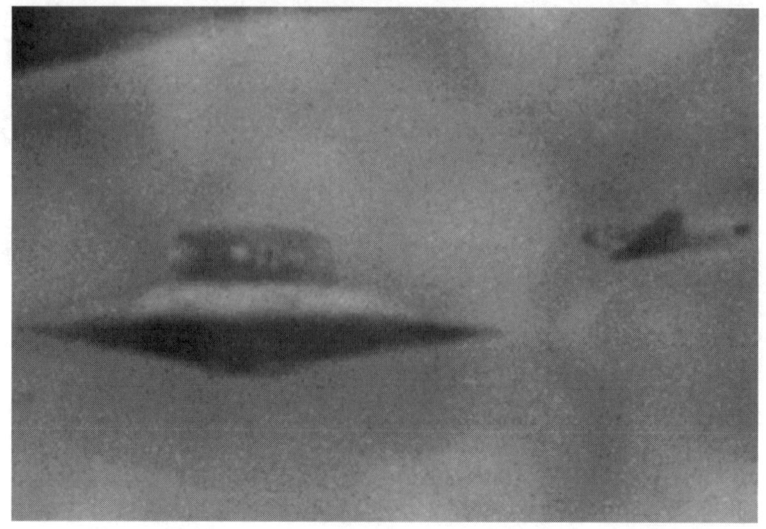

A photo of a Vril craft in flight with an Me-109 in the background.

BMW 028 was flown by the test pilot Joachim Roehlike at Peenemunde. The vertical rate of ascent was 800 meters per minute, it reached a height of 24,200 meters and in horizontal flight a speed of 2,200 km/h. It could also be driven with unconventional energy. But the helicopter never saw action since Peenemunde was bombed in 1944 and the subsequent move to Prague didn't work out either, because the Americans and the Russians occupied Prague before the flying machines were ready again.

In the secret archives of the SS the British and the Americans discovered during the occupation of Germany at the beginning of 1945—photographs of the Haunebu II and the Vril I crafts as well as of the Andromeda device. Due to President Truman's decision in March 1946 the war fleet command of the U.S. gave permission to collect material of the German high technology experiments.

Under the operation PAPERCLIP German scientists who had worked in secret were brought to the U.S. privately, among them VIKTOR SCHAUBERGER and WERNHER VON BRAUN.

A short summary of the developments that were meant to be produced in series: The first project was led by Prof. Dr. mg. W. 0. Schumann of the Technical University Munich.

Under his guidance seventeen disk-shaped flying machines with a diameter of 11.5 m were built, the so-called VRIL-1-Jager (Vril-1 fighters), that made 84 test flights. At least one VRIL-7 and one VRIL-7 large capacity craft apparently started from Brandenburg—after the whole test area had been blown up—towards Aldebaran with some of the Vril scientists and Vril lodge members.

The second project was run by the SS-W development group. Until the beginning of 1945 they had three different sizes of bell-shaped space gyros built: The Haunebu I, 25m diameter, two machines built that made 52 test flights (speed ca. 4,800 km/h). The Haunebu II, 32m diameter, seven machines built that made 106 test flights (speed ca. 6,000 km/h). The Haunebu II was already planned for series production. Tenders were asked from the Dornier

A photo of a Vril craft in flight very near a road and car.

and Junkers aircraft manufacturers, and at the end of March 1945 the decision was made in favor of Dornier. The official name for the heavy craft was to be DO-STRA (DOrnier STRAtospehric craft). The Haunebu III, 71m diameter, only one machine built that made at least 19 test flights (speed ca. 7,000 km/h).

The ANDROMEDA DEVICE existed on the drawing board, it was 139m long and had hangars for one Haunebu II, two Vril I's and two Vril II's.

There are documents showing that the VRIL 7 large capacity craft was used for secret, still earth-bound, missions after it was finished and test flown by the end of 1944:

1. A landing at the Mondsee in the Salzkammergut in Austria, with dives to test the pressure resistance of the hull.

2. Probably in March and April 1945 the VRIL 7 was stationed in the "Alpenfestung" {Alpine Fortress} for security and strategic reasons, from whence it flew to Spain to get important personalities who had fled there safely to South America and "NEUSCHWABENLAND" to the secret German bases erected there during the war.

From this fascinating list of craft and their statistics, information coming from the 1989 document dump in Britain that included all of the diagrams that we are seeing, we can gather that at the end of the war there were:

17 VRIL-1-Jager (Vril-1 fighters)
2 VRIL-7 (Seated crew of ?)
2 Haunebu I
7 Haunebu II (The Haunebu II seated 9 crew, could hold up to 20)
1 Haunebu III (The Haunebu II seated 32 people)
2? Andromeda Craft (Device)

The Andromeda craft, called a device in the Vril material on the Internet, was said to be a 139-meter-long tubular craft that, according to plans released, held one Haunebu II and two Vril craft. The above text says that there was four Vril craft in the Andromeda mother ship, so there is a discrepancy here. We know that the crew of a Haunebu II was up to nine people, and that it could ultimately hold 20 people, but we do not know what the crew of the Andromeda craft would be. We might surmise by calculating in this way: A crew of nine for the Haunebu II and another four crewmembers for the smaller Vril craft. If there was another 20 possible crewmembers we get a figure of 33 crewmembers.

This is an interesting number for a secret society of Teutonic Knights who believed that they were fighting a World War against an English-French Masonic society who believed that they were descended from the Knights Templar. The Templars, after their suppression, founded the first modern banks and courts at the Old Bailey in London's Old City. The underground stop here is Temple, and the original circular Templar church in London can be found here as well. The Teutonic Knights were a third branch of the Crusaders, with the Knights of St. John and the Knights Templar being the other two branches. The Nazi SS fancied themselves as the reincarnations of the Teutonic Knights of old—a group that was allied with the British and Templars, but ultimately opposed to the Russians and other Slavic forces.

In many ways World War II was about secret societies fighting for control of the world in the 1930s and 1940s. There was still time for the Germans and Japanese and their allies to grab more control

A photo of a Haunebu II craft on the ground.

A photograph of a Haunebu II craft on the ground.

of the world, and the Thule Society, the Vril Society and the SS all preached the message of an expansive Germany—a Germany that was all over the world, like the British Empire. But their nemesis was the British Empire, a Masonic-Templar establishment that had its own secret societies. It is important to remember that secret societies like the Masons, Knights of Columbus, and many others were all the rage until television came along in the 1940s and 1950s. The British were actually seen by Hess and other members of the Thule Society as allies, not enemies. However Churchill and the British military would not negotiate on any terms with Hitler's regime.

While the documents from the Third Reich released in 1989 do not state that the Andromeda craft was ever built, photos of the craft apparently exist. This seems to be the craft of the Adamski photos and many others. It seems difficult to believe that there was only one of these craft in existence, but that may be the case.

It seems quite possible, however, that a number of these craft were produced at the end of the war—and in the years after—in the various secret manufacturing facilities in Germany, Antarctica, South America, Greenland and elsewhere. The Germans were famous for shipping parts for various airplanes to all sorts of the world so they could be assembled when the time was right. Some of the stories even say that the Germans had the parts for long-range aircraft at the base(s) in Antarctica but they were never able to fly for the lack of certain parts.

Submarines That Can Fly

It is impossible to know just what occurred during the war with flying saucers and the technology involved. Many researchers believe that the Vril and Haunebu craft are real but that much of the material on the Internet is some form of disinformation, possibly aimed at neo-Nazi groups. Still, something like this was going on and even though Germany officially surrendered in May of 1945 it is painfully obvious with hindsight that portions of the Third Reich—overtly and covertly—moved their operations to South America and, in a sense, Antarctica. Also, we have seen that operations in Germany and other parts of Europe did not come to a complete halt either.

Since much of the Black Fleet, long-range aircraft, and now flying saucers, was under the control of the SS with its Black Sun symbol, with many of its members formerly of the Thule Society or the Vril Society, they did not surrender. Like any secret society that comes under attack and threatened with elimination, they did what they originally did—they become a secret organization again and operated accordingly. They were not destroyed or dissolved—they still functioned.

Stevens says that after the war ended the Nazi International-

A diagram of Vril craft in a launch shaft from an underground base.

A photograph of a Haunebu II in flight with a cannon on the underside.

Third Power set up scientific workshops in South America for the research and development of high technology. This included experimentation and production of flying discs. This coincided with the golden age of flying saucers in the late 1940s through the 1960s. Flying saucer accounts in and around South America abounded, and even today the attitude toward UFOs in all of the countries in South America is one of casual acceptance. The local newspapers and television stations have no problem reporting on UFO activity, and it has been a common topic of discussion since the end of WWII.

Stevens notes how a strange FBI report claimed a German inventor, living in South America, had discovered a new principle of "aerodynamics" and was trying to market this idea. His concept was not one based upon free energy in that he stated his method involved using motorcycle or auto engines. These engines produce movement of a spinning shaft, whose speed and torque can be harnessed for many uses but at its core is rotary motion.

Stevens says that the modern era of inquiry into the high technology of Nazi Germany only began in the middle to late 1980s and then only hesitantly. One of those first to delve in was the German writer O. Bergmann who published two short books[51] propounding the idea that later German flying discs were field propulsion powered and that there is absolutely no difference between a flying saucer and a flying U-boat at this point. He says they are exactly one and the same, and that these devices made their first appearance at the very end of the Second World War.

Stevens says that Bergmann bolstered his argument with the now familiar "Vril" and "Haunebu" diagrams, in fact he was the first to publish them, and with testimony from all over the world, covering decades of mystery submersibles and UFOs entering and leaving the water. A passage from one of Bergmann's books bears repeating:

> At sometime and someplace, at secret U-boat bases outside the German motherland, U-boats of the German Navy must have been weighed out, and, during the great withdrawal in April/May 1945, missing U-boats must have been equipped with new revolutionary technology and also must have been converted with electromagnetic propulsion. With this, those [U-boats] may, [have been] arranged with the same possibilities and technologies as the German flying discs [called UFOs].[51]

Bergmann goes on and on making this point. As an example, another passage:

A modern model for the Haunebu II craft.

A painting of a Haunebu II craft on the ground from one of the models.

UFOs or USOs are certainly not only observed within the seas and oceans and rivers, on the contrary also occasionally diving and surfacing on inland lakes, yes, even in ponds. We need not bother ourselves any more concerning the confusion on differing reports, if it is to be called only UFO, USO or still U-boat because [John] Keel had it completely correct, with the same electromagnetic propulsion it can be what it wants to be, the initial water craft would be able to operate in the air, as other flying craft, as well as the water. The two mediums, air and water, are interchangeable if one is fitted with this phenomenal propulsion, concerning which we have already more closely read in Hugin-Schrift: 'Geheime Wunderwaffen' Bd. III. They could maneuver mutually well in the air or water.[51]

Anti-Gravity or "Field Propulsion"

This anti-gravity or "field propulsion" is apparently a relatively simple technology that is basically electric in nature, rather than jet or rocket powered. Says Stevens about the technology:

> The first clues come from the two scientists who did a ground floor evaluation of Karl Schappeller, his home, his home area, witnesses and the Schappeller technology itself. Their conclusion was that the technology involved was first introduced by Tesla, and that Schappeller experimented exclusively with Tesla ideas. So does this mean we ought to be looking for clues to Haunebu secrets by looking at old Tesla stuff? Why not? It already seems that Tesla has been involved in the smaller type of field propulsion German flying disc.
>
> ...But from Schappeller this technology clearly found its way to the researchers at the Reichsarbeitsgemeinschaft (RAG), a governmental organization charged with energy independence for Germany at a time between the two World Wars. So the RAG was primarily not interested in levitation but nevertheless freely commented on this as being contained among the properties of the device they were building.

Steven then quotes from the Reichsarbeitsgemeinschaft:

> The new dynamic technology will enable, in the future, electric locomotives and automobiles to be made without expensive armatures and through circuit-connection to the atmospheric voltage-net to be pushed forward. Prerequisite is that, all things considered, the installation of many sufficient amplification facilities (centers) which transmit the specific "magnetic impulse" given from the Ur-machine to the dynamic globe-elements. Novel aircraft with magneto-static propulsion and steering, which are crash proof and safe from colliding with each other could be built for a fraction of the cost of today's aircraft—and without lengthy training of every person attending (the machines).

Stevens continues by discussing the curious globes that are often seen on the bottom, and sometimes the sides, of the craft:

> In conjunction with their discussion on energy production and levitation of aircraft the RAG describes how this all looks to the observer. They propose using seven globes, five surrounding one in the center with one positioned above this center globe. The five feed the center globe and when it reaches some point of saturation, specific magnetic impulses are sent to the seventh. This is what they describe as an Ur-machine. The reader may recall from depictions of the Haunebu machines that there are sometimes three globes visible and presumably one or two larger ones within the body of the craft. The three smaller globes may have an additional function in steering the craft but they also may interact with the globes in the body of the craft to produce magnetic reconnection and the lift producing positrons that result.
>
> Along with the Haunebu type of saucer, this technology would also be suitable for the cigar-shaped flying craft as well as the similarly shaped underwater, unidentified submerged objects. Just as would the two counter-rotating magnetic wheels of the previous homopolar generator inspired devices. For a submarine, the engine should probably be mounted in the front of the ship or perhaps one at the front and one at the rear for forwards and backwards motion. In other words William R. Lyne, all those old German sources, D. H. Haarmann and O. Bergmann are all probably correct, U-boats can fly, sometimes.

This last statement is profound, and as already stated, this electric field technology is essentially able to work underwater as it does in the air. Therefore, incredibly, with this technology the Germans could literally turn submarines into airships, hence the elongated, cigar-shaped UFOs, dark in color, that are commonly seen and even photographed starting immediately after WWII.

Where would the switchover from the old technology to the new

SS General Hans Kammler, center, with other officers.

one take place in order to turn these U-boats in airships? Probably built at the submarine base in Antarctica, and possibly at the bases in the Arctic. Supplies could be brought to the Antarctic base from Argentina and even from southern Africa. Most of the original food and other supplies would have come from Europe.

Though Germany was defeated, the secret SS laboratories, with their black U-boats, long-range aircraft, and even flying saucers, continued to repair older craft but also built new craft, mainly the flying disks, like the Haunebu. South America became a hotbed of UFO reports for decades and it would appear that many of these craft were assembled in the southern parts of South America, namely Argentina and Chile. Paraguay is another country that had a pro-German stance right up through the 1980s and today. A large ranch in western Paraguay could have all the privacy and support that might be needed to run a secret airbase. Curiously, the Bush family has a large ranch in western Paraguay. Large ranches with airstrips can be makeshift airbases for a short time, and both Lyndon Johnson and George W. Bush had long airstrips on their ranches, where even cargo planes and Air Force One landed.

The Mystery of Hans Kammler

Hans Kammler was a high-ranking Nazi and an SS-Obergruppenführer, the highest rank in the SS, and one of the most powerful men in the Third Reich right up until the end of the war, and probably beyond. Kammler joined the Nazi Party (NSDAP) in 1931 and held a variety of administrative positions after the Nazi government came to power in 1933, initially as head of the Aviation Ministry's building department. He joined the SS in May of 1933. In 1934, he was a councilor for the Reich's Interior Ministry.

Kammler was also charged with constructing facilities for various secret weapons projects, including manufacturing plants and test stands for the Messerschmitt Me 262 and V-2. Following the Allied bombing raids on Peenemünde in Operation Hydra, in August 1943, Kammler assumed responsibility for the construction of mass-production facilities for the V-2. He started moving these production facilities underground, which resulted in the Mittelwerk facility and its attendant concentration camp complex, Mittelbau-Dora, which housed slave labor for constructing the factory and working on the production lines.

SS General Hans Kammler in uniform during the war.

In 1944, Himmler convinced Adolf Hitler to put the V-2 project directly under SS control, and in August Kammler replaced Walter Dornberger as its director. From January 31, 1945, Kammler was head of all missile projects. During this time he was also partially answerable for the operational use of the V-2 against the Allies, until the moment the war front reached Germany's borders. As an SS officer, Kammler was the last person in Nazi Germany to be appointed to the rank of SS-Obergruppenführer. In March 1945, partially on the advice of Goebbels, Hitler gradually stripped Goering of several powers on aircraft support as well as maintenance and supply while transferring them to Kammler.

Kammler is listed as dead on May 9, 1945, the day that he disappeared and one day after the official surrender of the Third Reich. Kammler essentially vanished into thin air with numerous stories told in later years to explain his death or disappearance. He was said to have: committed suicide by taking a cyanide pill and then to have been secretly buried; been captured by US troops and brought to the US; continued to live in Austria or somewhere else in Europe; fled to South America. One might imagine that Kammler was on one of the flights to Tibet during or after the war.

Henry Stevens mentions that Kammler had probably been part of a planned coup to overthrow Hitler in what was known as Operation Avalon or Schwarze Adel, so he might have been considered to have been part of the "resistance" if one were willing to stretch the term.[46]

He also says that an eyewitness, a German engineer named Hans Rittermann, spoke with Kammler in his new position long after the war ended, and Kammler was not in hiding in South America but in Prague, sustained by both the East and West and working for both sides, the Americans and the Soviets.

Like Wernher von Braun, Kammler must have sent intermediaries ahead to cut a deal before his actual surrender and he must have had a plan to disclose some facet of the technology he was charged with developing as a bargaining tool. This might have been, for instance, sending someone ahead to divide the file in question into two parts says Stevens.

Kammler had worked and made his career in what is now the Czech Republic. Kammler spoke Czech. Prague is a world of its own, neither fully Slavic nor German, but an island in between. This was a place where both the German and Czech languages could be heard on the streets. Kammler was home! He must have been given comfortable surroundings, and a good allowance or salary.

Stevens says that Kammler, now back in Prague dealing with both the Americans and Soviets, must have had connections to Otto Skorzeny and the Rheinhard Gehlen Organization with its program of bringing important Nazi scientists to the US in Operation Paperclip. Stevens says that Kammler would have been able to get word out to his former associates in the SS and he quickly became part of the Nazi International, an extra-national group with interests all around the world.

Stevens says that Kammler supplied high technology to both the East and West in exchange for the technology that they had developed on many fronts. Says Stevens:

> The Nazi International would have been able to move around the planet in their flying discs and mystery U-boats at will. The Gehlen Org. would have given them a "heads up" to any problems and supplied them with insider, secret information about both the East and West. Gehlen, by some accounts, stoked the fires of the Cold War, inflating some intelligence to keep the confrontation going. This all worked together so well. In a way, Kammler had his Fourth Reich and he was Fuehrer.
>
> The axis of this "Fourth Reich" or Third Power as it should be known, was Madrid (Skorzeny), Munich (Gehlen), and probably the Prague/Pilsen area for Kammler—an area that he knew so well.
>
> The workshops were in South America, being supplied with sub-assemblies and parts for "Mimes Schmiede" most from the USA. Money was no problem since Dr. Hjalmar Schacht was alive and well and the Third Power was probably making money on its own on an ongoing basis. This trade in Nazi science and technology and the relationship with both the East and West probably had its ups and downs but went on until the early-mid-1970s when Kammler died (1972) and Skorzeny died (1975). This was about thirty years after World War Two and the older generation must have been

A photo of Otto Skorzeny as an SS officer.

> considering retirement. With the coming of the 1980's, there must have been some sort of change of leadership and perhaps new goals and outlooks.

So apparently Kammler lived for decades after the war, and probably even took a flying saucer ride or two, but always had to live in hiding. Like Rheinhard Gehlen, Otto Skorzeny was able to live his life in partial exile in Madrid and did not have to go into hiding.

Skorzeny was a tall, Austrian-born SS-Obersturmbannführer (lieutenant colonel) during World War II. During the war, he was involved in many operations, including the removal from power of Hungarian Regent Miklós Horthy and the famous rescue mission that freed Benito Mussolini from captivity in a mountaintop castle.

Skorzeny was held in Germany by the Allies for two years but he escaped from an internment camp in July of 1948 with the help of three SS officers disguised as American guards. He hid out on a Bavarian farm for 18 months, then spent time in Paris and Salzburg before eventually settling in Francoist Spain where he was given asylum. Though Skorzeny was unable to travel to certain countries, like Germany, Austria, Britain or the United States, he was essentially a free man and could travel freely with a Spanish passport to many countries, including Ireland, Switzerland and all of South America.

Starting in 1950 Skorzeny traveled frequently between Spain and Argentina, where he acted as an advisor to President Juan Perón and as a bodyguard for Eva Perón, while fostering an ambition for the "Fourth Reich" to be centered in Latin America.

In 1953 he became a military advisor to Egyptian President Mohammed Naguib and recruited a staff of former SS and Wehrmacht officers to train the Egyptian Army, staying on to advise President Gamal Abdel Nasser.

In 1962, Skorzeny was allegedly recruited by the Mossad and conducted operations for the agency against missile scientists in Egypt.

In the 1960s, Skorzeny set up an early group of for-hire mercenaries called the Paladin Group, which he envisioned as "an international directorship of strategic assault personnel [that would] straddle the watershed between paramilitary operations carried out by troops in uniform and the political warfare which is conducted by civilian agents." They were based near Alicante, Spain, and specialized in arming and training guerrillas. Some of its operatives were recruited by the Spanish Interior Ministry to wage a clandestine war against the Basque terrorist group ETA.

Skorzeny had homes in Ireland and Majorca, as well as in Madrid. He may have had property in Argentina, though he probably didn't really need it. His friends had plenty of land and even flying saucer factories. Officially, Skorzeny died of lung cancer on July 5, 1975 in Madrid at the age of 67. His funeral in Madrid was attended by a number of former Nazis as well as Mossad officers. Some writers have suggested he lived on past this funeral, and finished his final days at an estate in Florida where his daughter lives to this day.

We can see that the trio of Rheinhard Gellen in Austria, Skorzeny in Madrid and Kammler in Prague would be a strong axis of evil to control a worldwide empire of clandestine businesses and factories. But we never learn who the chief officer in South America was. Perhaps it was Skorzeny who was constantly moving between Spain and Argentina in the early years after the war. Or, perhaps our mystery man—the Dr. Evil of South America—was someone else, someone who remains anonymous today, though it likely that such a

Otto Skorzeny, on left, with Juan Peron, center, in Buenos Aires in the early 1950s.

person is dead by now, having died in the 1970s or 1980s. It has been said that Juan Peron sold Martin Bormann ten thousand Argentine passports in 1947, so perhaps it was Bormann who was "our man in Argentina," though he is officially listed as killing himself with a suicide pill in 1945. We just don't really know if Bormann had any real hand in what was going in South America.

But anyway, apparently Skorzeny was in constant contact with Kammler and Gehlen and would then fly to Argentina—on a commercial flight—and meet with his many Nazi contacts in that country. From Argentina Skorzeny may have visited the base in Antarctica, probably in a Haunebu.

Without support from South America the base in Antarctica could not function for an extended period of time. Let us look now at some of the strange goings on in South America.

UFOs and Secret Cities of South America

A curious story is told by Jim Keith, in his book *Casebook on the Men in Black*,[31] about a Spaniard who was visiting Venezuela and had a strange encounter. Keith says that Guillermo Arguello de la Motta, a Spanish physician, was visiting friends at a small town near Caracas, Venezuela in July of 1971 when he observed two men in black suits, red ties and black berets get out of a red Mustang sedan and then stand there by the car. Says Arguello in Keith's book:

> The two men stood there waiting for about five minutes and then began to put on orange-colored belts, talking together animatedly in the meantime. Suddenly a shiny object appeared in the sky. It rapidly descended and then stopped at a height of about 60 centimeters from the ground. It was circular, bell-shaped underneath, and with a turret on the upper part. Its width could have been about 30 meters. What surprised us most of all about it [were] the rapid changes of color, from orange to blue and then to white. When it halted, floating in the air, it rotated through almost 180 degrees. Suddenly a small parabolic staircase came down from the base of it, which enabled the two men from the Mustang car to enter the saucer with ease. When the staircase had been drawn in again, the craft dipped slightly towards its left side

> and then, following an inclined flight path, vanished into the sky at an impressive speed. The machine was of course definitely no helicopter. It was totally silent, and its shape was something totally unknown, i.e., not conventional.[31]

The witness never did see who was inside the craft, but the two men in black suits and black berets—Men in Black—were simply waiting for the craft as if it were a city bus, put on orange belts and hopped on board. Ride in a flying saucer, anyone? Stevens says that during the 1950s, 60s and 70s it was necessary for these secret SS officers and Project Paperclip Nazis to travel in this manner, rather than risk travelling through major airports with forged passports. Many of the former SS officers who had served in the war were getting quite elderly by the 1970s and ultimately these last officers came in from the cold, so to speak.

Ultimately, says Stevens, they became part of the secret space program—a space program that involved Antarctica in a major way.

Jim Keith describes another curious encounter in Latin America, this one near Mexico City on May 3, 1975. Keith says that a private aircraft pilot named Carlos de los Santos nearly collided with three disk-shaped craft while traveling in a small plane near Mexico's large capital city, an encounter that was confirmed by radar operators at the airport. The next week de los Santos claimed that he was driving to a television station to discuss his close encounter with the three disks, an appointment he missed, when he was forced to stop on the highway by two black Ford Galaxie limousines. Says Keith:

> Four men in black suits approached de los Santos and warned him in Spanish in a "mechanical tone" to be quiet about the incident "if you value your life and your family's too." A month later on his way to visit with American scientist and ufologist J. Allen Hynek, de los Santos met an MIB on the hotel steps. The MIB shoved him and made a statement leading de los Santos to believe that he had been under observation. Again, de los Santos missed his appointment.[31]

In that same year, 1975, an Argentine anthropologist and UFO

A photograph of Cerro Uritorco in northern Argentina.

investigator named Anton Ponce de Leon described the strange occurrences that happened to him while investigating some UFOs around a small Peruvian town named Sicuani. Sicuani is a town on the main highway—a two-lane paved road—between Cuzco and Lake Titicaca. It is a major truck stop town with numerous gas stations, restaurants, hotels and shops—but still not a particularly large town. Ponce de Leon describes his investigation thusly:

> In the year 1975 there was much talk about UFO sightings in Sicuani, particularly at night, and the peasants were talking about something very worrisome to them, that they were hearing noises under the ground as if a machine were working there, according to their expression.[31]

Keith says that Ponce de Leon met with a newspaper reporter from Lima, Peru, who worked for the paper there called *Ultima Hora*. This reporter told Ponce de Leon that he had film of some flying saucers taken about a week before in Capilla del Monte, Argentina, a northern Argentina town many hundreds of miles to the south. This area is famous for a series of UFO disk sightings in the early to mid 1970s. Says Wikipedia about the town and the UFO phenomena centered on the nearby mountain called Cerro Uritorco:

> Capilla del Monte is a small city in the northeastern part of the province of Córdoba, Argentina, located by the Sierras Chicas mountain chain, in the northern end of the Punilla Valley. It has about 11,281 inhabitants as per the 2010 census.
>
> The main tourist attraction in the area is the Cerro Uritorco, a small mountain only three kilometers from the city, famed around Argentina as a center of alleged paranormal phenomena and UFO sightings. The city also features the tall El Cajón Dam and its large reservoir. Capilla del Monte was founded on October 30, 1585. Its name means "Mount's Chapel" in Spanish.

Capilla del Monte is a hotbed of UFO disk activity, which apparently continues to this day. The unnamed reporter from the *Ultima Hora* newspaper had photos of the disk craft in Argentina and was now at the Peruvian town of Sicuani with his undeveloped film investigating other UFO sightings when he met Anton Ponce

A screen shot of a Haunebu at Cerro Uritorco from a television newscast.

A poster for Cerro Uritorco in northern Argentina promoting UFO sightings.

de Leon.

The photographer told Ponce de Leon that two men in black suits had come to his hotel when he was not there and told the clerk that they were his friends and asked to wait in the room for the photographer's return. Instead they ransacked his room, apparently looking for the film, and then left. The photographer had actually given the film to a friend in town to develop.

Later he met the men in black suits at the hotel lobby where they demanded his film. He told them the film had already been picked up and they left, but he found that his room continued to be entered and papers that were on a table were taken. He eventually fled the town in a taxi to Cuzco and then flew to Lima where the photos were ultimately published in an issue of *Ultima Hora*.[31]

This is just a sample of the strange happenings in South America that went on for decades, with numerous UFO sightings and even the strange story of noises coming from underground like there was machinery at work. Was it possibly a tunnel-boring machine beneath the city? If so, was this some postwar operation being conducted, not by the government of Peru, but of some Third Power of ex-Nazis operating out of Antarctica and Argentina? Is there a secret UFO base somewhere near the northern Argentina town of Capilla del Monte? It would seem so.

The Final End of the Third Power and the Dark Fleet

The evidence seems clear that the second World War did not

come to an end with the surrender of Germany and Japan. It has been pointed out earlier that the SS never actually surrendered at the end of the war, and some claim that Japan never officially surrendered either. Unlike Germany, which was overrun by Russian, British and American forces, Japan was never overrun by the Allied forces, but rather two of its major cities were "nuked" and thus Japan capitulated to the Allies. At the time of capitulation Japan essentially still occupied all of the territory that it had taken during the war. It took nearly a month for Japan to pull its forces back to Japan, leaving a massive power vacuum in many areas formerly under its control. Many of these countries, such as North and South Korea, are still dealing with this sudden end to the fighting in the Asian theater; these countries began to have their own fights inside their own countries—divisions promoted by the Americans on one side and the Russians on the other.

The Black Fleet with its secret SS labs and secret cities in South America—and even Antarctica—managed to last for decades as a Third Power that controlled whole countries in South America at times. We can easily put Argentina, Chile and Paraguay into this category, and other countries such as Brazil, Bolivia and Peru were heavily influenced by this Third Power that apparently used secret saucer bases in the mountainous areas of these countries.

But the big question is whether this Third Power was assimilated into the secret space program and whether the Nazi base in Antarctica came under the control of the American-Russian secret space program.

Stevens says that the Third Power ultimately came to a truce and agreement of sorts with the United States military during the second term of President Ronald Reagan. Stevens says that about the same time as his famous "Mr. Gorbachev, tear down this wall" speech on June 12, 1987, Reagan had made a side trip to an obscure town in far Western Germany, a town called Bitburg. At a small ceremony there Reagan placed flowers at the graves of the German soldiers who had died there defending their country from the invading Western Allies near the end of the war. There are no Americans buried at Bitburg, but rather there are a large amount of Waffen SS (the military branch of the Nazi Party's SS organization) soldiers buried at the small cemetery. These SS officers had fought against the British and Americans at the end of war.

Why would President Ronald Reagan do this? Stevens suggests that this was a formal integration of the former Waffen SS and its many tentacles into the US military's secret space program. President Reagan also ordered that all Pershing missiles based in Germany be removed. Reagan had a very pro-German foreign policy while at the same time offering overtures to the Russians, who also wanted the Pershing missiles removed from Germany.

The United States was at this time given some special technology by the Third Power, Stevens claims, including maser and laser technologies that the United States didn't have. They also gave the Americans klystron technology that can be used to produce positrons for space travel. We might think that some of the space technologies were developed at the secret SS labs in the Arctic, Antarctica and South America—and possibly even Tibet. Earlier we discussed briefly that the SS still maintained laboratories in Western Germany well into the 1950s. No such secret SS labs would have been possible in Eastern Germany, it would seem.

Says Stevens about klystron anti-gravity technology in which beams of positrons are projected above or below a discoid craft:

> Then let us suppose a flying craft powered by positrons generated via klystron technology. How would this look? Well, how about mounting the klystron-positron projectors as swivel heads, mounted on the wing tips of a flying triangle? At the point where the three beams converge, that is the point into which the flying triangle moves, pushed and pulled by this super Biefield-Brown Effect. At low power, for instance, with the three beams converging over the center of the craft's gravity, it would just hang there, motionless. With variations in power and focus, any type of three-dimensional movement is possible.
>
> If this were the technology the Reagan Administration had received, it would account, temporally, for the appearance of flying triangles in Belgium only a couple years after "payment" had taken place. Likewise, beyond Belgium this new shape seemed to be the predominant UFO shape from that time to the present.[46]

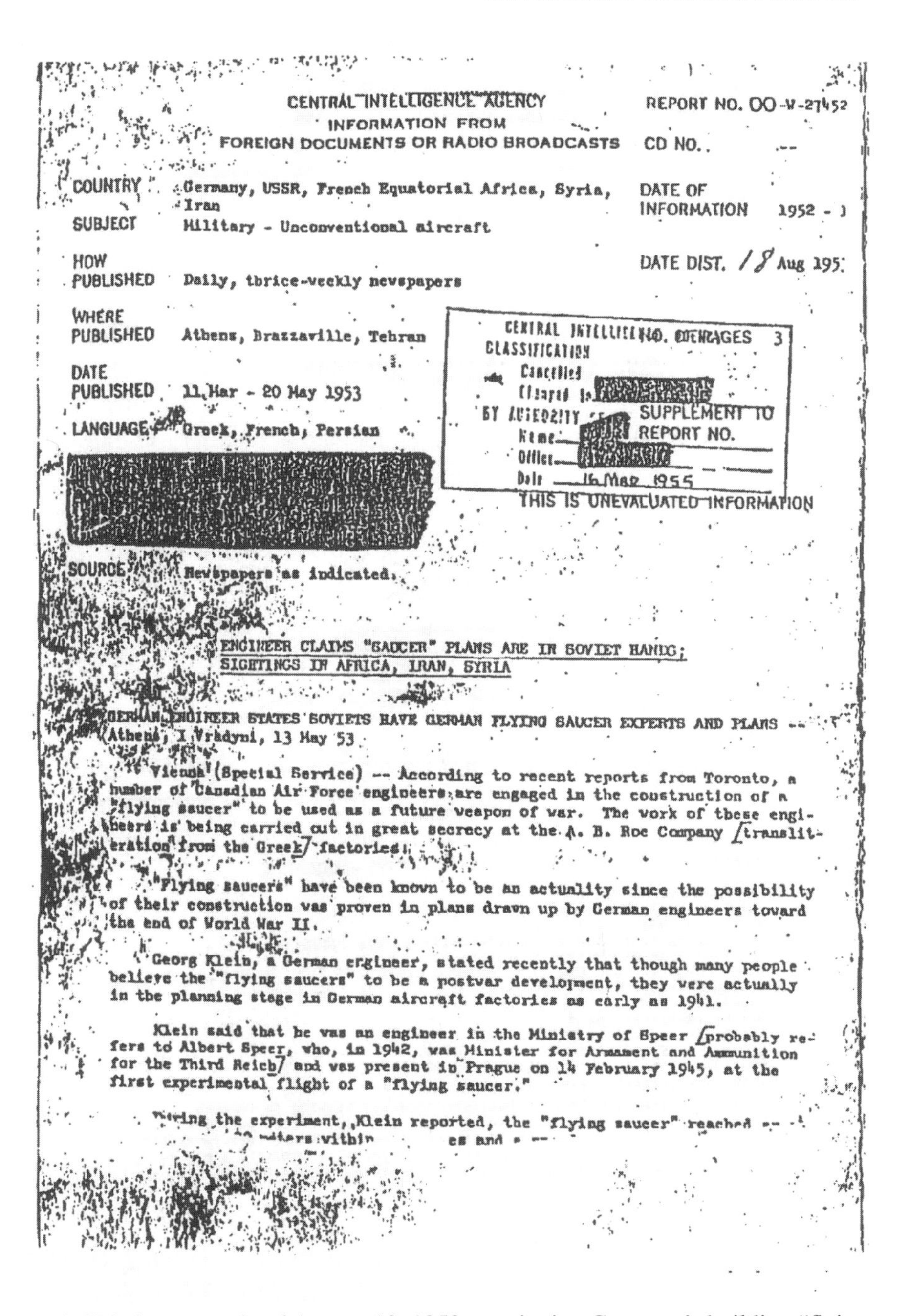

CENTRAL INTELLIGENCE AGENCY
INFORMATION FROM
FOREIGN DOCUMENTS OR RADIO BROADCASTS

REPORT NO. OO-W-27452
CD NO.

COUNTRY Germany, USSR, French Equatorial Africa, Syria, Iran

SUBJECT Military - Unconventional aircraft

HOW PUBLISHED Daily, thrice-weekly newspapers

WHERE PUBLISHED Athens, Brazzaville, Tehran

DATE PUBLISHED 11 Mar - 20 May 1953

LANGUAGE Greek, French, Persian

DATE OF INFORMATION 1952 - 1

DATE DIST. 18 Aug 195

NO. OF PAGES 3

CENTRAL INTELLIGENCE CLASSIFICATION
Cancelled
Changed to [illegible]
BY AUTHORITY OF
Name [illegible]
Office [illegible]
Date 16 Mar 1955

SUPPLEMENT TO REPORT NO.

THIS IS UNEVALUATED INFORMATION

SOURCE Newspapers as indicated.

ENGINEER CLAIMS "SAUCER" PLANS ARE IN SOVIET HANDS; SIGHTINGS IN AFRICA, IRAN, SYRIA

GERMAN ENGINEER STATES SOVIETS HAVE GERMAN FLYING SAUCER EXPERTS AND PLANS -- Athens, I Vradyni, 13 May 53

Vienna (Special Service) -- According to recent reports from Toronto, a number of Canadian Air Force engineers are engaged in the construction of a "flying saucer" to be used as a future weapon of war. The work of these engineers is being carried out in great secrecy at the A. B. Roe Company [transliteration from the Greek] factories.

"Flying saucers" have been known to be an actuality since the possibility of their construction was proven in plans drawn up by German engineers toward the end of World War II.

Georg Klein, a German engineer, stated recently that though many people believe the "flying saucers" to be a postwar development, they were actually in the planning stage in German aircraft factories as early as 1941.

Klein said that he was an engineer in the Ministry of Speer [probably refers to Albert Speer, who, in 1942, was Minister for Armament and Ammunition for the Third Reich] and was present in Prague on 14 February 1945, at the first experimental flight of a "flying saucer."

During the experiment, Klein reported, the "flying saucer" reached ... meters within ... es and ...

A CIA document dated August 18, 1952 mentioning Germany's building "flying saucers" as early as 1941. US Navy intelligence probably knew about this, but they had never told the CIA. Even the director of the CIA kept this a secret.

This was also a time period when the final officers of the Third Reich, including Operation Paperclip Nazis like Wernher von Braun were dying of old age. Says Henry Stevens:

> And this time-line ties into what was happening internally within the Third Power itself. The Third Power remained intact until sometime around the 1990's, in the technological sense. Wilhelm Landig had kept in contact with expatriate Reich insiders who informed him of the situation. The reader will remember the "Ruestungsesoteriker," the armament esoterics, the ultimate geeks who kept the postwar Nazi high technology, including the field propulsion flying discs in good operation. They worked out of these secret bases and through a network supplying parts and know-how toward the perpetuation of these technological wonders which resided in huge underground caverns connected to the outside world by tunnel-entrances. Well, it seems even they had limits. All machines eventually wear out.[46]

Stevens then tells us that the German SS officer and novelist Wilhelm Landig commented on the postwar situation through an informant, Heiner Gehring, who managed a peek into Landig's private files:

> Actually, after the war, there were efforts made to build a military-technical power outside of Germany. The Base 211 in Neuschwabenland, which really was established, was certainly given up after some time. Besides the immunity situation it may have also been the dropping of the atom bomb of the USA during the Geophysical Year 1954 as a reason for the handing over of the base. Garrison and materiel of the Base 211 were sent to South America. Likewise, more German U-boat bases were available.
>
> German flying discs, so the Landig particulars let it be known, are still warehoused in South America but perhaps may not be flight worthy any more. Their propulsion was unconventional. The development of the "Haunebus"

> through the "Vril" Society did, in collaboration with the SS, actually take place. Certainly some of those topics that are found in the video-films in circulation are partially completely humbug. False are the reports concerning moon and Mars flights of German flying discs, the equipment was not developed for such effort. In the USA and USSR, attempts after the Second World War were made to replicate flying discs. This replica has, on propulsion, failed, which can not be reconstructed.[46]

So Landig admits that the Antarctic base, as well as the bases in the Arctic, were probably abandoned after they were "nuked" in 1954. However, it would seem that other bases in Antarctica continued to operate well into the 1960s and the secret cities in South America may well still be active to this day. Likewise, the secret airbase in Tibet is likely to have been abandoned by the early 1960s as the Communist Chinese extended their power into even the remotest corners of Tibet. Landig mentions that the craft are warehoused in South America, and may be unusable. This does not seem to be the case, as South America continues to be a hotbed of UFO stories and sightings and it would seem that craft of different sorts, even newer models, are still being flown in South America as well as Antarctica.

Landig is also probably incorrect when he says that Americans were unable to reconstruct the field effect propulsion that propelled the Haunebu and other craft. It seems likely that the Americans, who had allegedly teleported a battleship in the Philadelphia Experiment, moved forward with their own secret anti-gravity projects immediately after the war. It is interesting to note as well that Landig dismisses any flights to the Moon or Mars during the war or afterward, stating that the craft are for terrestrial flights. No mention is made of "making submarines fly." Can they fly to the Moon? Many researchers believe that they can.

Says Stevens on the ageing of the generals of the Third Power:

> Dr. Hans Kammler was said to have died in 1972 according to Hans Rittermann. Otto Skorzeny died in 1975. Reinhard Gehlen died in 1979. Afterward, certainly, a

younger generation of Nazis took over management of the Third Power. By the early-mid 1980's the Cold War was moving into its fourth decade with no end in sight or in contemplation in the minds of anyone on planet earth. Even the younger men running the Third Power must have been thinking about their own retirement as they were doubtless already in their 60s. It is no wonder they were more willing to divulge their secrets at this time. They were not sought criminals or marked men in any sort of way. They were free to fly first-class on commercial jets, they had no need of flying discs at this point.

The point is this postwar Nazi sub-culture existed in parallel along side of both the Eastern and Western blocks of the Cold War, interacting and influencing both of them in ways which we are now only beginning to understand. Just as Farraday and Tesla postulated a self-inducting homopolar generator, these Nazis self-inducted the Cold War to a greater or lesser extent. And in some ways, especially in terms of technology, this postwar Nazi subculture was the focus of the conflict. And yet on the other hand, in terms of their middle position regarding communications, they were the glue which kept the Cold War from overheating. Was the Third Power was a balance-point, a fulcrum? Maybe this is overstating things but in the more narrow scope, especially in Europe and with high technology, it must have been important.

The Third Power was as much a child of the Cold War as were the Eastern and Western blocs and when one power got off this teeter-totter, shifting the world order, the existence of the Third Power as an active world-player became redundant. Now, twenty years later, their flying saucers rest in cold mountain storage, their U-boats rust in still-secret places, and their microfilm slowly deteriorates in hidden caches.

Today the Third Power is not trying to take over the world or even become a Fourth Reich. The Third Power was a concept for its members, and lives within friendly government officials and organizations of several countries as well as corporations and financial entities that owed a measure of their success to this Third Power. It also lives in the organizations who owe their roots or at least a greater part of their nourishment to Nazi ideas. This includes the Vril Society and the Karoteckia both of which, I am told by reliable sources, still exist. It also lives in the technology of those times, some of which must still be secret.[46]

The discussion of Nazi flying saucers resurfaced in August of 2018 when a series of declassified CIA files were reported by news outlets around the world including the *Daily Star* of the UK. The newspaper said on August 6, 2018 that the CIA files were about interviews with a German engineer named George Klein from between March 11 and May 20, 1952. Said the *Daily Star*:

> In these interviews Klein claimed that the Nazis had a flying saucer that was capable of reaching heights of 12,400 meters in three minutes—with speeds of up to 2,500 mph. Nazi flying saucers appear twice in the CIA's trove of documents as part of their investigations into UFOs. The Klein testimony was published in newspapers in Greece, Iran and the Congo.
>
> Klein said he was an engineer in the Ministry of Speer (i.e.: Albert Speer, Reich Minister of Armaments and War Production for Nazi Germany) and was present in Prague on February 14, 1945, at the first experimental flight of a flying saucer. Klein claims the Third Reich actually successfully carried out a test of their "flying saucer" in Prague on Valentine's Day, 1945, only months before the Czech capital was liberated by the Soviet Union's advancing Red Army.
>
> Exceeding 2,500 mph (Mach 3) would make the Nazi saucer almost twice as fast as the state-of-the-art F-35 warplane being rolled out in the US and UK during World War II. Klein claimed the saucer could take off vertically like a helicopter, and had been in development since 1941.
>
> The CIA document alleges the saucers were constructed at the same slave-labor driven factories which made the dreaded V2 rockets. Nazi engineers were reportedly

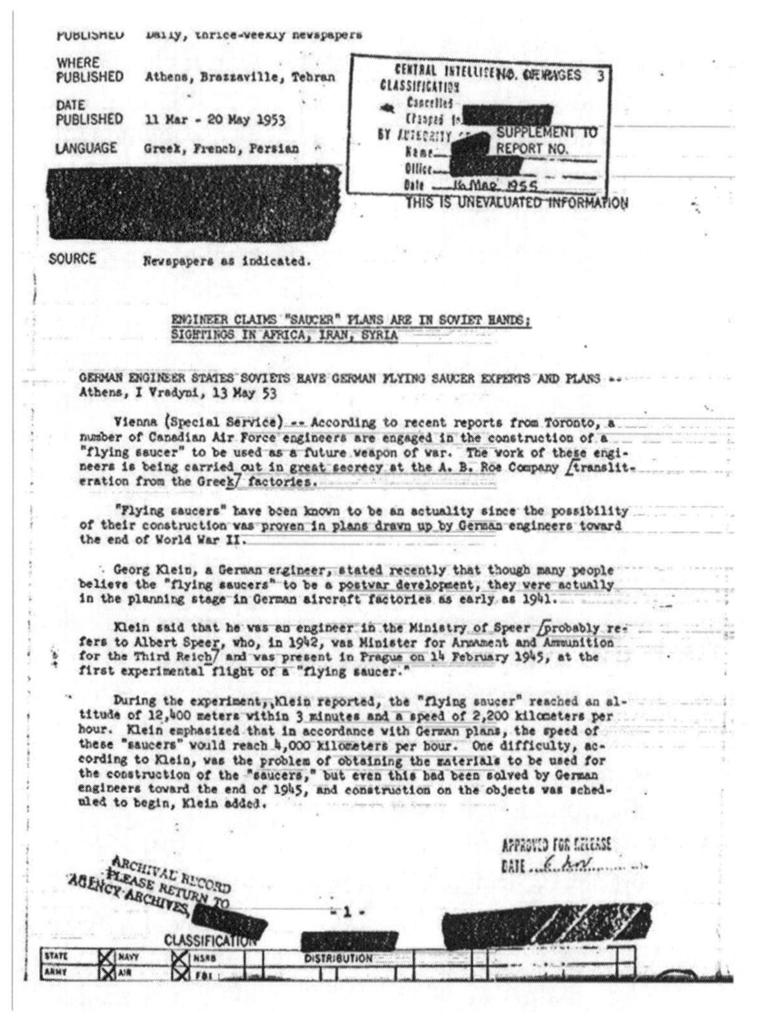

PUBLISHED Daily, thrice-weekly newspapers

WHERE PUBLISHED Athens, Brazzaville, Tehran

DATE PUBLISHED 11 Mar - 20 May 1953

LANGUAGE Greek, French, Persian

CENTRAL INTELLIGENCE CLASSIFICATION Cancelled Changed to BY AUTHORITY OF Name Office Date 16 MAR 1955

NO. OF PAGES 3

SUPPLEMENT TO REPORT NO.

THIS IS UNEVALUATED INFORMATION

SOURCE Newspapers as indicated.

ENGINEER CLAIMS "SAUCER" PLANS ARE IN SOVIET HANDS;
SIGHTINGS IN AFRICA, IRAN, SYRIA

GERMAN ENGINEER STATES SOVIETS HAVE GERMAN FLYING SAUCER EXPERTS AND PLANS -- Athens, I Vradyni, 13 May 53

Vienna (Special Service) -- According to recent reports from Toronto, a number of Canadian Air Force engineers are engaged in the construction of a "flying saucer" to be used as a future weapon of war. The work of these engineers is being carried out in great secrecy at the A. B. Roe Company [transliteration from the Greek] factories.

"Flying saucers" have been known to be an actuality since the possibility of their construction was proven in plans drawn up by German engineers toward the end of World War II.

Georg Klein, a German engineer, stated recently that though many people believe the "flying saucers" to be a postwar development, they were actually in the planning stage in German aircraft factories as early as 1941.

Klein said that he was an engineer in the Ministry of Speer [probably refers to Albert Speer, who, in 1942, was Minister for Armament and Ammunition for the Third Reich] and was present in Prague on 14 February 1945, at the first experimental flight of a "flying saucer."

During the experiment, Klein reported, the "flying saucer" reached an altitude of 12,400 meters within 3 minutes and a speed of 2,200 kilometers per hour. Klein emphasized that in accordance with German plans, the speed of these "saucers" would reach 4,000 kilometers per hour. One difficulty, according to Klein, was the problem of obtaining the materials to be used for the construction of the "saucers," but even this had been solved by German engineers toward the end of 1945, and construction on the objects was scheduled to begin, Klein added.

APPROVED FOR RELEASE DATE 6 Nov

ARCHIVAL RECORD PLEASE RETURN TO AGENCY ARCHIVES

- 1 -

CLASSIFICATION

STATE	X	NAVY	X	NSRB		DISTRIBUTION			
ARMY	X	AIR	X	FBI					

A 1953 CIA document discussing German flying saucer technology.

evacuated from Prague as the Red Army bore down upon them. Klein claims one team failed to be notified of the order to escape—and they were captured by the Soviets.

The CIA file reads: "Klein stated recently that though many people believe 'flying saucers to be a postwar development, they were actually in the planning stage in German aircraft factories as early as 1941." "Klein was of the opinion that the 'saucers' are at present being constructed in accordance with German technical principles and expressed the belief that they will constitute serious competition to jet-

> propelled airplanes." Klein claimed the saucers were being built by the Russians, but no known Soviet saucers ever materialized.
>
> Claims about the Nazis' advanced technology are often tied to alleged links between Hitler and the occult. Conspiracy theorists claim that remnants of the Third Reich fled to South America under the guidance of SS commander Hans Kammler. The Nazis are claimed to have continued testing their experimental weapons from secret bases in the Antarctic. UFO sightings from the era are alleged to be secret tests of experimental technology by the Nazis—and the US and Soviets. Theories persist to this day about advanced Reich technology—and still cause controversy.

Indeed, it would seem that this Third Power does still exist in South America, given the numerous UFO reports that come out of that mysterious continent. The craft can't all be from other planets. Rumors have existed for decades that the wealthy Tesla student Guglielmo Marconi had faked his death in 1937 and had moved with over a hundred European scientists to a secret jungle location in Venezuela or Colombia and set up a space center based on Tesla technology. This secret space base in South America was the plot of the 1979 James Bond film *Moonraker*.

What now of the flying submarines? Are the former German bases in Antarctica just empty bunkers waiting for intrepid explorers to find their dark caverns and light them up one more time? Were they destroyed by the Americans in the 1970s? Or perhaps they continue to be fully operational—manned by the men and women of the secret space force. This is the most powerful space force on the planet, and many people believe that it is being commanded from Antarctica.

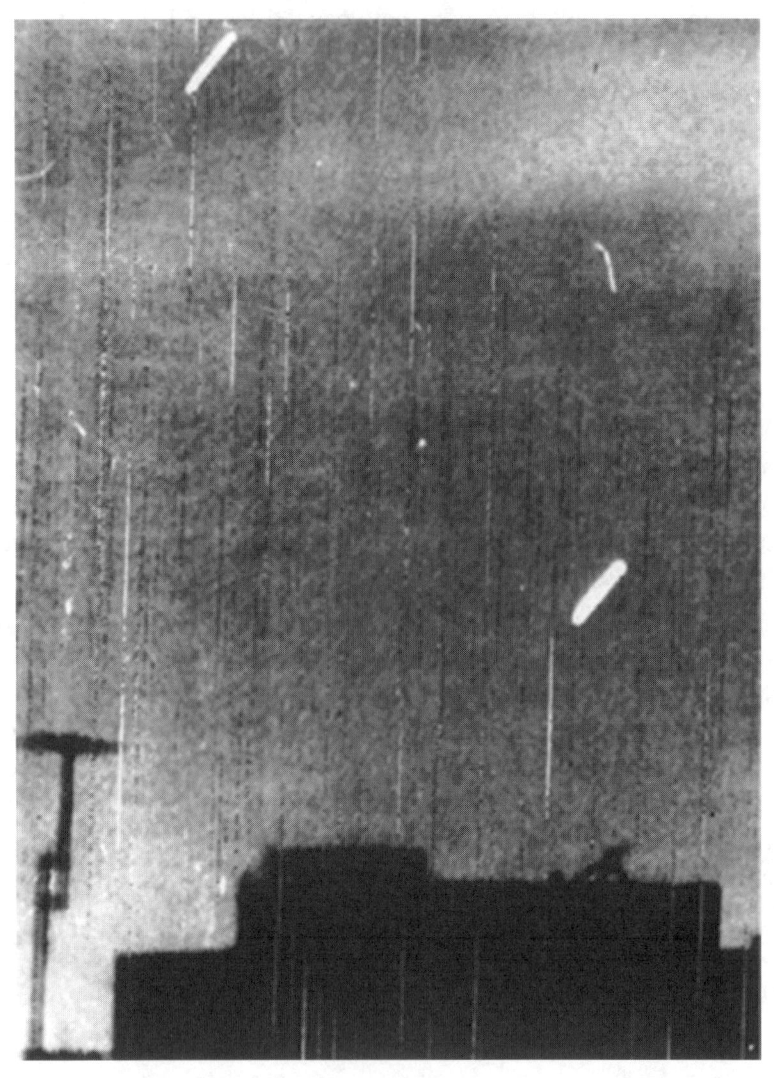

A photograph of cigar-shaped craft over Buenos Aires, Argentina in 1965.

Chapter Eight

Project Horizon and Solar Warden

When you're lost in Juarez, when its Easter time, too,
And your gravity fails and negativity can't pull you through…
—Bob Dylan, *Just Like Tom Thumb's Blues*

You can't fight a war in space by yourself. You can't have a race to the Moon by yourself. You can't have an arms race by yourself. It is a simple fact that one needs an opponent to have an arms race or a space race. During WWII our opponents were armed and dangerous and the Axis powers needed to be defeated at great cost and sacrifice by Americans and others. But with the war over and Germany and Japan defeated and occupied, who were the bad guys to arm against?

President Eisenhower would later warn America in his last televised speech about the dangers of the military-industrial complex. A major portion of the military-industrial complex involves aerospace and aerospace-related products. Nikola Tesla had died in New York City during the war (1943) but Albert Einstein continued to work with the US Navy after the war—he was allegedly a consultant on the Philadelphia Experiment—and famously sent a letter to President Truman in 1946 that suggested that the United States form a partnership with the Soviet Union to bring stability and peace to a fractured world.

In a sense, such a combined force for planetary good was promoted in the 1960s by the *Man from UNCLE* television series in which Russian and American agents battled a third power known as THRUSH. As we saw in the last chapter, the THRUSH organization was a thinly veiled international Nazi network complete with mad scientists, arms dealers and wealthy but shady businessmen. During this time period the activities of the Odessa and Ratlines groups were beginning to surface and certain Nazis that had escaped to South America were captured and brought to Israel or Germany for trial. The "myth" of Hitler and other top Nazis having escaped to South America and other places had a resurgence in the media, and well known Nazis and former Nazis like Otto Skorzeny and Wernher von Braun were making the news.

But, despite the efforts of Hollywood and *The Man from UNCLE*, America would get immersed in its worldwide rivalry with the Russians—and a Cold War and race to the Moon would commence. This was not just a race to the Moon, but a race to dominate space and ultimately the solar system. Hopefully no one had gotten there before them…

The Mysterious Black Knight Satellite

Yes, there would be a race to the Moon, and the Russians shot forward with the launching of *Sputnik* in 1957, followed soon after by the American *Telstar* satellite. A mysterious satellite was already in orbit at this time, the "Black Knight" satellite. Some think that this satellite was launched sometime in the 1940s, perhaps by the Germans at the very end of the war. And as we have seen, we are not really sure when this war actually ended.

In 1954, UFO researcher Donald Keyhoe told newspapers that the United States Air Force had reported that two satellites orbiting Earth had been detected. At that time, no country was known to have the technology to launch a satellite. One of these satellites was supposedly the "Black Knight" satellite that is in a polar orbit. The second satellite was also of unknown origin.

Perhaps the Black Knight satellite was launched from Antarctica at some point after the official end of WWII, possibly in 1945 or 1946 when German operations in Antarctica, South America, Greenland and Africa (Spanish Sahara and the Canary Islands) were still active. With a telecommunications satellite in a polar orbit around the earth the surviving Nazis would have had a superior worldwide communications system a decade ahead of the Allies.

With the launch of the Black Knight, the Nazis of the Third Power were able to step up their game of chess with the Allies and go into space with a satellite or two. The Black Knight satellite was probably launched with a modified A-4 rocket brought in parts via submarine to Antarctica. It is also possible that the satellite was launched from the Peenemunde rocket site in Germany in the closing days of the war.

Still, the idea that Antarctica may have had a role in the launching of the Black Knight satellite is intriguing. Was Germany able to fly manned missions into space around this time as well? Popular legends, and even movies, have suggested that the Nazis were able fly the Hanuebu flying saucer to the Moon, and even Mars. Others have claimed the Vril and Haunebu saucers were not designed for space travel and are limited to travel within the earth's atmosphere.

Along these lines, one has to ask: could the "flying submarines" that were claimed to have existed just after the end of the war have flown to the Moon and back? If so, this would be the very beginning of a secret Moon base, one that may be connected with the early secret base in Antarctica.

The establishment of a secret base on the Moon would be the primary objective of any secret space program. Such a secret space program—typically run by the military—would involve the launching of rockets from military bases loaded with parts and people for the clandestine lunar facility. Indeed, beginning in the early 1950s the US Army began drafting up plans for exactly such a secret military base on the Moon. It was called Project Horizon and a full report was published in 1959 but kept secret for many decades. From what can be learned from the declassified papers, Project Horizon was an astonishing proposal to have a military base on the Moon by 1966. Yes, 1966!

Project Horizon: The Secret Base on the Moon

Project Horizon was a 1950s study to determine the feasibility of constructing a scientific/military base on the Moon. During this period their was no NASA. Rather there were three separate militaries in the USA with three separate space programs: the US Department of the Army, Department of the Navy, and Department of the Air Force.

In 1958 the Air Force released a memo on its Lunex Plan, and on June 8, 1959, a group at the Army Ballistic Missile Agency

(ABMA) produced for the Army a report titled "Project Horizon, A U.S. Army Study for the Establishment of a Lunar Military Outpost." The project proposal states the requirements as follows:

> The lunar outpost is required to develop and protect potential United States interests on the moon; to develop techniques in moon-based surveillance of the earth and space, in communications relay, and in operations on the surface of the moon; to serve as a base for exploration of the moon, for further exploration into space and for military operations on the moon if required; and to support scientific investigations on the moon.

The permanent outpost was predicted to be required for national security "as soon as possible," and to cost $6 billion. The projected operational date with twelve soldiers on the Moon was December 1966. Officially, Project Horizon never progressed past

A 1959 artist's drawing for the US Army Moon base as part of Project Horizon.

the feasibility stage. It was said to have been rejected by President Dwight Eisenhower and the primary responsibility for America's space program was supposedly transferred to the civilian agency that was just being created, NASA. But was Project Horizon really cancelled? Why would it be? Apparently the funding was there for the Army's Moon base and it is normally the function of military programs to build complicated "forward" bases, whether they be in Antarctica, or on the Moon or Mars.

Indeed, it seems very clear that both the Army and Navy continued their own space programs for many decades and still have space programs to this day, now under a unified command structure called Space Force. It is obvious that the Army and the Navy continued with their Moon base plans and probably met their goal of having a 12-man crew on the Moon by late 1966.

Space Transportation System

From what little we know about Project Horizon from the released documents, it was estimated to require 147 early Saturn A-1 rocket launches to loft spacecraft components for assembly in low Earth orbit at a "spent-tank space station." A spent-tank space station would utilize a spent rocket stage as part of its construction. A liquid-propellant rocket primarily consists of two large, airtight propellant tanks; it was realized that the tanks could be retrofitted into the living quarters of a space station.

From this newly constructed space station—a secret military installation in orbit around the Earth—smaller space vehicles could journey to the Moon and back. The project papers said that a lunar landing-and-return vehicle launched on a Saturn A-2 would have shuttled up to 16 astronauts at a time to the base and back.

So apparently they would launch a smaller Saturn A-2 rocket from the space station and it would journey to the Moon. This rocket would have the lunar landing-and-return vehicle at the nose in a similar configuration as the later Apollo missions. However, this Army lunar landing-and-return vehicle could hold up to 18 passengers, while the Apollo capsule could only hold three.

Says the Wikipedia entry for Project Horizon:

> Rocket-vehicle energy requirements would have limited

the location of the base to an area of 20 degrees latitude/longitude on the Moon, from ~20° N, ~20° W to ~20° S, ~20° E. Within this area, the Project selected three particular sites:

northern part of Sinus Aestuum, near the Eratosthenes crater

southern part of Sinus Aestuum near Sinus Medii

southwest coast of Mare Imbrium, just north of the Montes Apenninus mountains

Construction

1964: 40 Saturn launches.

January 1965: Cargo delivery to the Moon would begin.

April 1965: The first manned landing by two men. The build-up and construction phase would continue without interruption until the outpost was ready.

November 1966: Outpost manned by a task force of 12 men.

This program would have required a total of 61 Saturn A-1 and 88 Saturn A-2 launches up to November 1966. During this period the rockets would transport some 220 tons of useful cargo to the Moon.

December 1966 through 1967: First operational year of the lunar outpost, with a total of 64 launches scheduled. These would result in an additional 120 tons of useful cargo.

Defenses

The base would be defended against Soviet overland attack by man-fired weapons:

Unguided Davy Crockett rockets with low-yield nuclear warheads

Conventional Claymore mines modified to puncture pressure suits

Layout

The basic building block for the outpost would be cylindrical metal tanks, 10 feet (3.0 m) in diameter and 20 feet (6.1 m) in length. Two nuclear reactors would be located

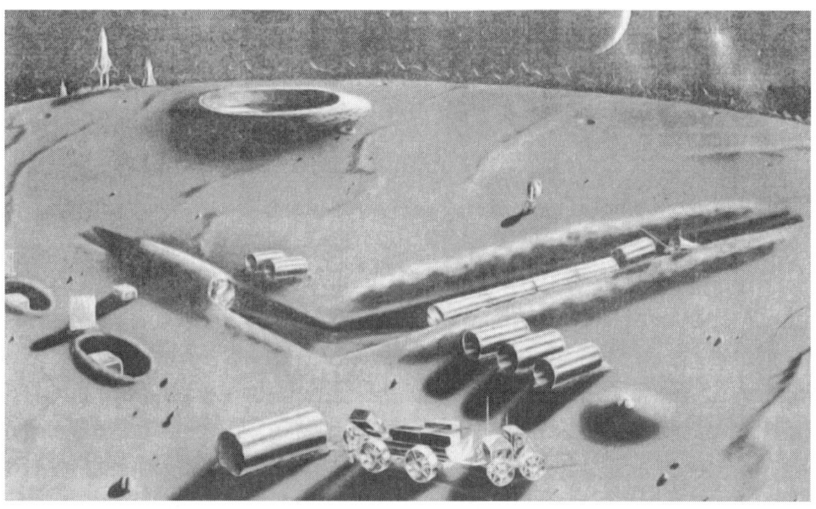

A 1959 artist's drawing for the US Army Moon base as part of Project Horizon.

in pits for shielding, and would provide power for operation of the preliminary quarters and for the equipment used in the construction of the permanent facility. Empty cargo and propellant containers would be assembled and used for storage of bulk supplies, weapons, and life essentials.

Two types of surface vehicles would be used, one for lifting, digging, and scraping, another for more extended distance trips needed for hauling, reconnaissance and rescue.

A lightweight parabolic antenna erected near the main quarters would provide communications with Earth. At the conclusion of the construction phase the original construction camp quarters would be converted to a bio-science and physics-science laboratory.

In his book, *Hidden Agenda*,[14] Mike Bara says:

The actual design and layout of the base was to be an "L" shaped configuration, consisting of buried or partially buried cylindrical metal tanks approximately 10 feet in diameter and 20 feet in length. Two nuclear reactors would be transported to the Moon and located in pits to provide shielding and power for the operation of the living quarters and for the construction of the permanent facility. Empty cargo and propellant containers would be assembled and

> used for storage of bulk supplies, weapons, and life support equipment like oxygen and water.
>
> ...The proposal was taken seriously enough that Wernher von Braun, then the head of ABMA, appointed Heinz-Hermann Koelle (another Nazi) to head the project team at Redstone Arsenal in Alabama.

So, here we have a perfectly viable and carefully-planned Lunar base, one that even had some Project Paperclip Nazi scientists working on it. But it never happened—or did it? Was this project really defunded? Maybe they went right ahead as planned and built their Moon base by 1966, then what? Might it have grown into a much larger base with other satellite bases at other sites on the Moon? Did the Navy and Air Force build similar bases on the Moon—bases that were independently run from the Army's bases? Are all these bases, including even a Russian base or two, now part of an integrated network of operational bases on the Moon? It would make perfect sense if missions to Mars were launched from one of these bases on the Moon, rather than from Earth. Was this von Braun's Mars Project come to fruition—one that would have

Fig. II-6. Typical Lunar Construction Vehicle

A 1959 artist's drawing for Project Horizon of a construction vehicle.

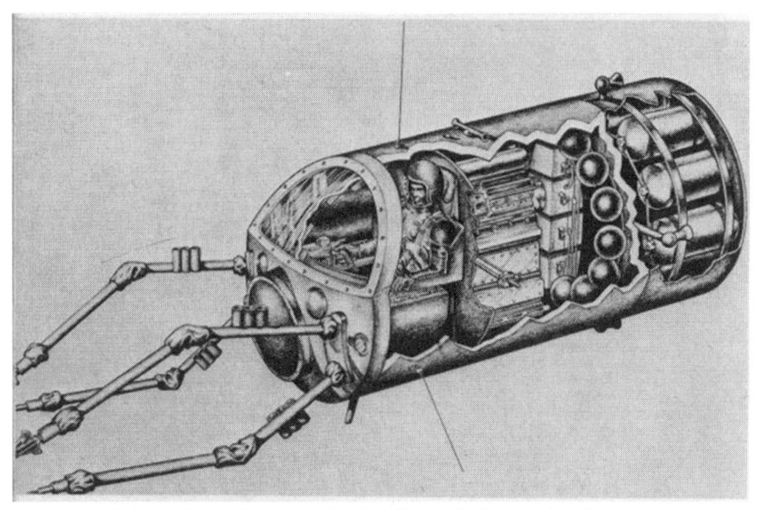

A 1960 artist's drawing for Project Horizon of a Space Tug.

rockets blasting off from the Moon to Mars?

So, to recap Project Horizon, its stated goal was to: a) build a space station in orbit around the Earth capable of launching rockets to the Moon; and b) build a base on the Moon that could house 12 personnel.

Not only do we have the Army's proposal for a lunar base, but the Air Force wanted one too. They called their Moon base the Lunex Project.

The Lunex Project

The Lunex Project was a secret US Air Force plan begun in 1958 for a crewed lunar landing prior to the Apollo Program. As the Lunex Project proceeded, it put forth a plan in 1961 to construct a

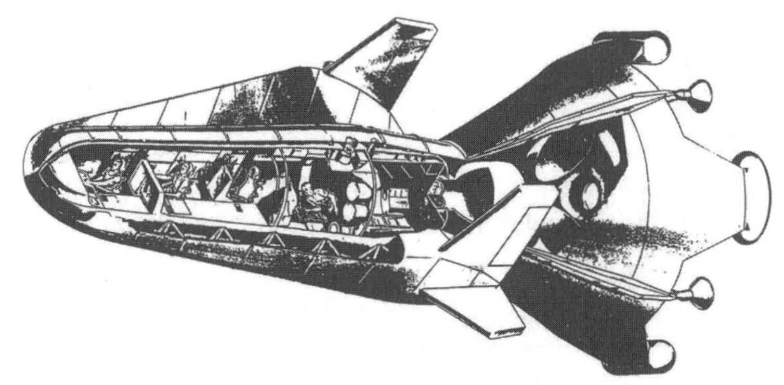

A drawing from the US Air Force's 1963 secret Lunex plan of the landing vehicle.

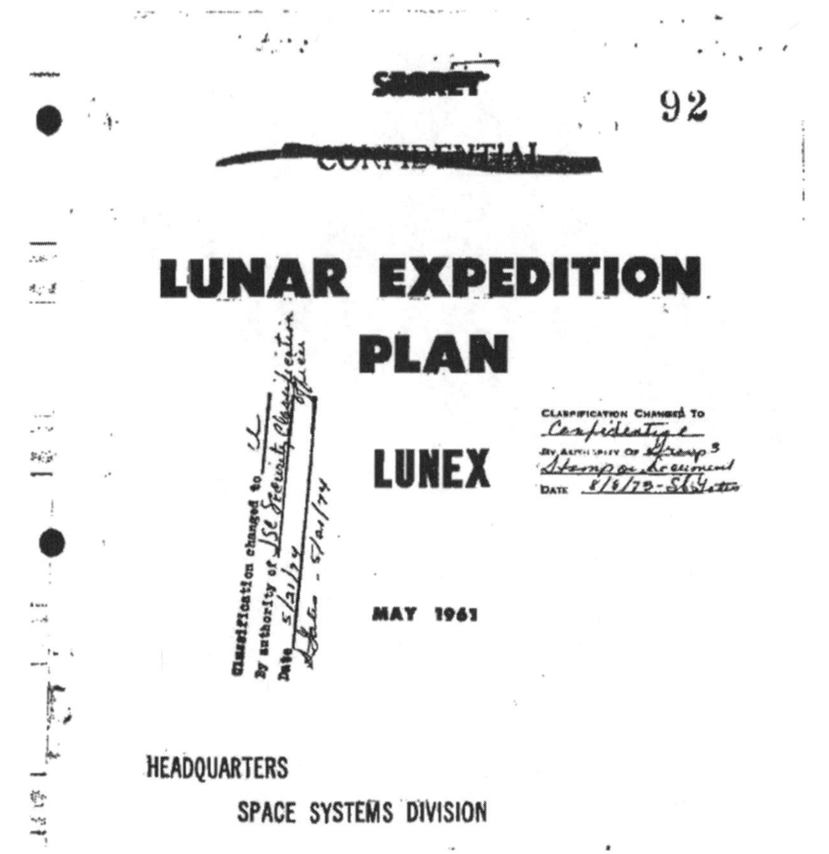

SECRET

92

CONFIDENTIAL

LUNAR EXPEDITION PLAN

LUNEX

Classification Changed To Confidential

By Authority Of Group 3

Date

Classification changed to

By authority of

Date

MAY 1961

HEADQUARTERS

SPACE SYSTEMS DIVISION

The cover page for the US Air Force's 1963 secret plan to put a base on the Moon.

21-person underground Air Force base on the Moon by 1968 at a total cost of $7.5 billion.

The plan was delivered four days after Kennedy's May 1961 speech directing NASA to aim for the Moon, seemingly a last-ditch attempt by the Air Force Ballistic Missile Division to stay involved in lunar exploration. It would use a six-million-pound-thrust cryogenic launcher for a "direct" ascent to the Moon. A crew of three would have a ten-day journey to the Moon and back inside a "lifting body" that would act as both lunar lander and reentry vehicle.

The primary distinction between the later Apollo missions and Lunex was the orbital rendezvous maneuver. The Lunex vehicle, composed of a landing module and a lifting body return/reentry module, would land the entire vehicle and all astronauts on the surface, whereas the final Apollo mission involved a separate ascent module leaving the command module and service module connected in lunar orbit with a single astronaut. The original plan for Apollo was for direct ascent, similar to Lunex.

According to documents that were finally released after being secret for decades, the Lunex Lunar Lander had a crew of three and the diagrams show two extra seats in the back. The length of the craft was 53 feet (16.16 meters) with a maximum diameter of 25 feet (7.62 meters) and it weighed 134,000 pounds.

The documents said that the selection of base sites was to be made by automated probes, with the Kepler crater being a location that they had looked into. The Kepler Moon crater is most notable for the prominent ray system that covers the surrounding mare. The rays extend for well over 300 kilometers, overlapping the rays from other craters. The outer wall of the crater is not quite circular, and possesses a slightly polygonal form while the interior walls of the crater are slumped and slightly terraced, descending to an uneven floor and a minor central rise. This might be a good crater for a

A 1967 photo of Kepler Crater from the Moon Orbiter 4 mission.

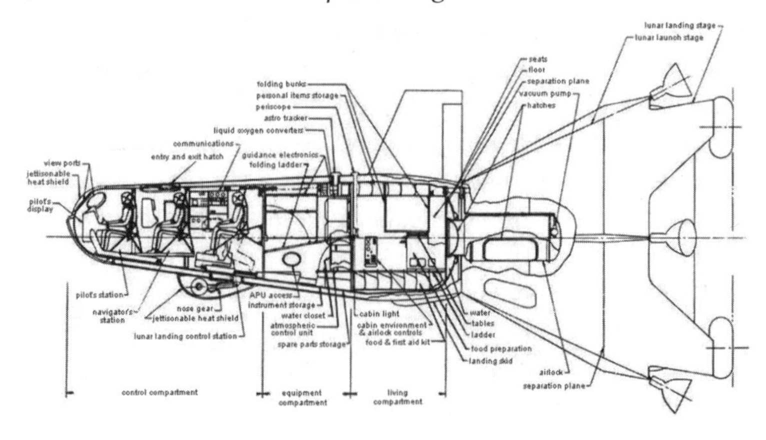

A drawing from the US Air Force's 1963 secret Lunex plan of the landing vehicle

Moon base.

Lunex planned to make its first lunar landing and return in 1967, in order to beat the Soviets and demonstrate conclusively that America could win future international competition in technology with the USSR. The Air Force felt that no achievement short of a lunar landing would have the required historical significance.

Three milestones were set for Lunex:

1965: recovery of a crewed reentry vehicle

1966: crewed circumlunar flight

1967: crewed lunar landing and return

After 1968, a Permanently Crewed Lunar Expedition was planned.

One of the admitted problems was the development of the lunar landing stage, which would have to make a precision landing tail-first on rocket thrust: something never previously tested. The crew would then make a departure from the Moon in the same craft, returning directly to Earth or Earth orbit.

A pdf of some of the documents related to the Lunex Project is available on the Internet if one looks carefully. Aside from the above information there are also several interesting drawings. One is of the Lunex lander craft itself, another is of a curious launch pad next to a cliff, another is of a "Lunar Fusion Reactor," and another is an illustration of what the Lunex base would look like.

The Lunex craft is interesting but there does not seem to be anything special about it. It is larger than the LEM used in the

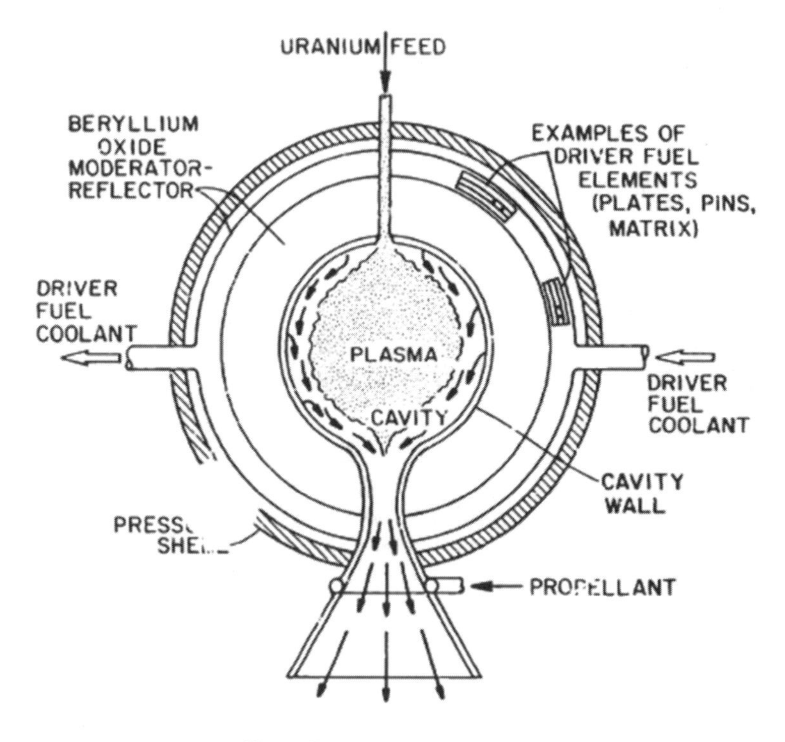

Figure 10. Gaseous Core Reactor Concept

A drawing from the US Air Force's 1963 secret Lunex plan of the fusion reactor.

Apollo missions, holding up to five or six persons. The Lunar Fusion Reactor is pretty interesting, with a beryllium oxide "moderator reflector" and uranium feed. The Lunex base was to have two of these fusion reactors.

The other diagrams are even more interesting, with several drawings of the launch pad, and what they show is somewhat surprising. One drawing is of the proposed Lunex Launch Complex from a distance. It shows a large body of water, an ocean or a large lake, with launch pads set up against cliffs that are near the water. Ramps are shown leading up to the top of the cliffs and storage and manufacturing buildings are at a safe distance from the launch site and are connected by roads, ramps or tunnels.

The other drawing shows a close-up of the Lunex Launch Complex. It shows a boat in the water approaching a series of two locks. These locks would bring a ship to the very base of the launch

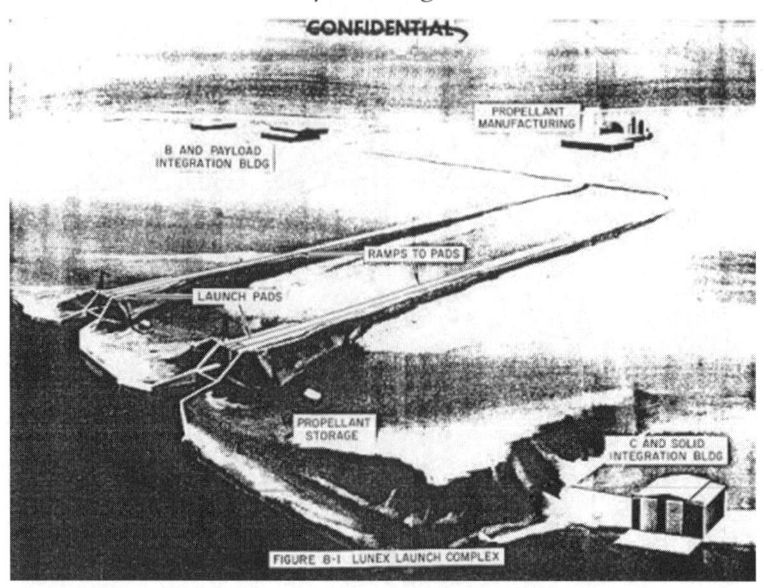

A drawing from the US Air Force's 1963 secret Lunex plan of the launch facilities.

site. This base of the launch site is also presumably reached by tunnels and elevators, but these are not shown. A rocket has been assembled on the launch pad and a crane is depicted on the top of the launch site, which seems to be a man-carved mountain and cliff, presumably out of granite. It is unlike any rocket launch site

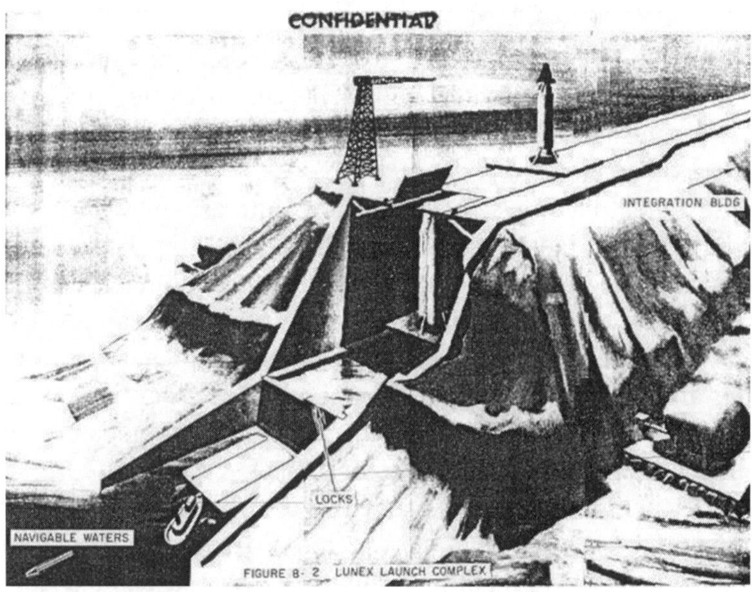

A close up drawing from the secret Lunex plan of the launch facilities.

depicted before.

The question that immediately leapt to my mind was, where is this launch site? Was it to be built in Antarctica? Was it ever built? Even if the Air Force and the Lunex Project never did go to the Moon, did they build this launching pad for other purposes? I cannot help thinking that this Lunex Launch Complex was built. And for that matter, the Air Force may well have built its base on the Moon.

And that brings us to the nuclear powered lunar base itself. The drawing provided looks like a bad photocopy of an actual photograph, but nevertheless a number of details can be seen. One sees antennas on buildings and several large hangar doors. Like Project Horizon, the Lunex base consisted of tubular modules that were welded together in sections, and after a few sections, there is a corner and the station turns at a right angle. This happens again after a few sections, and so on. What is curious here is that this tends to make the lunar base spiral out from its core section in a shape that looks very similar to the Nazi swastika.

Curiously, the 2012 Finnish-German-Australian film *Iron Sky* features a very similar lunar base, and this one is definitely meant to be in the shape of a swastika. *Iron Sky* tells the story of a group

A drawing of the Lunex Moon base with camouflaged buildings and right angles.

The lunar base built as a swastika in the film *Iron Sky*.

of Nazi Germans who fled to the Moon in 1945, where they built a space fleet to return in 2018 and conquer Earth. The movie is a comedy-action film and an obvious satire on the Nazis and their flying saucers. The Haunebu craft are featured in the movie as well.

The *Iron Sky* Moon base is fascinating to look at and one has to wonder if the set designers knew about the Lunex base as they planned their movie or whether it was just a coincidence. Also, is there really a base on the Moon that looks like a swastika? According to the Lunex documents, the answer is probably yes.

One other curious aspect to the Lunex Moon base diagram is that portions of it are camouflaged with mounds of soil. Other things such as the antennas are not camouflaged, but the main buildings and sections of tunnels have some obvious camouflaging around them. One has to wonder why this would be? Sure, the Army and the rest of the military like to camouflage things, we know, but why camouflage something on the Moon? One might actually think that they would want their lunar base to be as visible as possible. But apparently they wanted a hidden base on the Moon.

Was this camouflaging to keep the Russians from spotting our base on the lunar surface and bombing it? Maybe the camouflage was to hide the base from aliens—extraterrestrials that might already be using the Moon themselves. Or, incredibly, were they hiding the lunar base from Nazis—Nazis in Antarctica?

Even more startling, it is possible that the camouflage on the Lunex Moon complex was meant to hide the base from the American public, as well as the public at large. This is the astounding conclusion that we ultimately come to when it comes to any military base on the Moon—if it exists, it is meant to be hidden from the public and kept a secret except from the most closely guarded group of people, mostly in the military.

How do we know this? We know this because the military tells us, through our government officials (who are as ignorant of the subject as most people), that there are no military bases on the Moon, not Russian ones either. Instead they point out that NASA is planning to put a civilian international base on the Moon sometime in the future. Either this is indeed the case and all of these top secret projects were never acted upon, or secrets have got to be secrets.

One has to wonder, given the widely reported comments of the Apollo 11 astronauts that "they got here first" and that craft and structures could be seen, if the astronauts were merely seeing Army, Air Force or Navy Moon bases that were above their top secret clearance level.

The Apollo missions, for all their controversy, may have been just what they said they said they were, and the astronauts were just as surprised as anyone to find unusual things on the Moon. The NASA missions do not preclude the possibility of the military—all three major branches—having manned bases on the Moon.

A 1963 drawing from Boeing of two different Lunar Exploration Systems (LES).

And that brings us to our next subject, which is a mystery about the US Navy. We know that the US Army had a project to go to the Moon called Horizon. We also know that the Air Force had a lunar project called Lunex. But we do not know the name of the US Navy's lunar project—one they must certainly have had. Or do we? Perhaps the name of the Navy's secret Moon project is Solar Warden.

Solar Warden: The Navy's Secret Space Fleet

A secret space fleet code named Solar Warden was first reported in Britain by the news media when a computer hacker with Aspergers syndrome named Gary McKinnon was "arrested" by the British police in 2002. The basic story is told by British UFO investigator and journalist Darren Perks (who sometimes speaks in the third person) in an article that appeared in the huffingtonpost.co.uk on November 7, 2012. Said the article:

> … since 1980, a secret space fleet code named "Solar Warden" has been in operation unknown to the public that allegedly includes key world governments. While conducting an FOI (freedom of information) request with the DOD (department of defense) in 2010, I had a much unexpected response by email from them which read:
>
> "About an hour ago I spoke to a NASA rep who confirmed this was their program and that it was terminated by then President Obama. He also informed me that it was not a joint program with the DOD. The NASA rep informed me that you should be directed to the Johnson Space Center FOIA Manager."
>
> The program not only operates classified under the US Government but also under the United Nations authority. So you might be wondering how do I know this information? Well there are a few people and many others that have tried hard to find out the truth, and have succeeded by leaked information or simply asking questions and have government departments slip up and give away information freely, just like what happened when Darren Perks asked the DOD. One notable contributor is Gary McKinnon.
>
> When Gary McKinnon hacked into U.S. Space Command

computers several years ago and learned of the existence of "non-terrestrial officers" and "fleet-to-fleet transfers" and a secret program called "Solar Warden," he was charged by the Bush Justice Department with having committed "the biggest military computer hack of all time", and stood to face prison time of up to 70 years after extradition from UK. But trying earnest McKinnon in open court would involve his testifying to the above-classified facts, and his attorney would be able to subpoena government officers to testify under oath about the Navy's Space Fleet. To date the extradition of McKinnon to the U.S. has gone nowhere.

McKinnon also found out about the ships or craft within Solar Warden. It is said that there are approx eight cigar-shaped motherships (each longer than two football fields end-to-end) and 43 small scout ships. The Solar Warden Space Fleet operates under the US Naval Network and Space Operations Command (NNSOC). There are approximately 300 personnel involved at that facility, with the figure rising.

Solar Warden is said to be made up from U.S. aerospace Black Projects contractors, but with some contributions of parts and systems by Canada, United Kingdom, Italy, Austria, Russia, and Australia. It is also said that the program is tested and operated from secret military bases such as Area 51 in Nevada, USA.

So should we just write this off as utter nonsense?

No we shouldn't and as time goes on the truth will slowly come out. Many people around the world are now witnessing craft moving around in the skies and sub space that completely defy gravity. Whether they are part of the Solar Warden secret program, military experimental aircraft or not, thousands of people know what they see.

In my view Solar Warden is very real and a very strong possibility. So no, I don't think we should rule it out as complete nonsense.

Yes, it's a conspiracy because of all the hype and controversy surrounding the facts and information about the program.

Sensitive is an understatement. This program would

change the world and our views on space exploration and travel, so no wonder that it would be kept a big 'secret.'

We should all keep it in the back of our minds… for now at least!

According to this article, largely containing information from Gary McKinnon, we learn that the Solar Warden fleet consists of eight motherships, 43 smaller scout ships and over 300 military personnel, including officers. It is also worth noting that Russia is part of the consortium that also includes Australia, Canada, Italy, Austria and the UK. Notice that neither France nor New Zealand are listed. Nor is Germany. This is a curious collection of countries in some way, considering that New Zealand is part of the Five Eyes intelligence group with the UK, Australia, Canada and the US. Also, it might be noted that the 1989 document dump to Ralf Ettl with the photos and plans of the Haunebu and Vril craft is thought to have come from Austria.

An earlier mention of Solar Warden was on the Open Minds Forum in early 2006, though it came anonymously:

> We have a space fleet, which is code named 'Solar Warden.' There were, as of 2005, eight ships, an equivalent to aircraft carriers and forty-three 'protectors,' which are space planes. One was lost recently to an accident in Mars' orbit while it was attempting to re-supply the multinational colony within Mars. This base was established in 1964 by American and Soviet teamwork. Not everything is, as it seems.
>
> We have visited all the planets in our solar system, at a distance of course, except Mercury. We have landed on Pluto and a few moons. These ships contain personnel from many countries and have sworn an oath to the World Government. The technology came from back engineering alien-disc wreckage and at times with alien assistance.

The above claims are separate from what Gary McKinnon has said he found on computers at NASA and the Pentagon. Solar Warden does seem to be an international effort of a sort, and they may well have a base on Mars by now. One has to wonder if they have already been to Pluto and Neptune; this may just be disinformation. Gary McKinnon, however, is not disinformation, and Solar Warden was mentioned in British newspapers and other media starting in 2003 because of McKinnon's arrest.

But who is Gary McKinnon? Many Americans have never heard of him, but he is famous across the pond in Great Britain and something of a celebrity.

The Amazing Hacker Named Gary McKinnon

Gary McKinnon.

Gary McKinnon has his own Wikipedia page and from it we learn that McKinnon was born on February 10, 1966 and is a:

> …Scottish systems administrator and hacker who was accused in 2002 of perpetrating the "biggest military computer hack of all time," although McKinnon himself states that he was merely looking for evidence of free energy suppression and a cover-up of UFO activity and other technologies potentially useful to the public.
>
> On 16 October 2012, after a series of legal proceedings in Britain, Home Secretary Theresa May blocked extradition to the United States.
>
> McKinnon was accused of hacking into ninety-seven United States military and NASA computers over a thirteen-month period between February 2001 and March 2002, at the house of his girlfriend's aunt in London, using the name "Solo."

According to Wikipedia, US authorities stated he deleted critical files from operating systems, which shut down the United States Army's Military District of Washington network of two thousand computers for twenty-four hours. McKinnon also posted a notice on the military's website: "Your security is crap." After the September 11 attacks in 2001, McKinnon deleted weapons logs at the Earle Naval Weapons Station, rendering its network of three hundred computers inoperable and paralyzing munitions supply deliveries for the US Navy's Atlantic Fleet. The Pentagon also accused McKinnon of copying data, account files and passwords onto his own computer. US authorities stated that the cost of tracking and correcting the problems he caused was over $700,000.

McKinnon did admit leaving a threat on one computer:

> US foreign policy is akin to Government-sponsored terrorism these days… It was not a mistake that there was a huge security stand down on September 11 last year… I am SOLO. I will continue to disrupt at the highest levels…

US authorities stated that McKinnon was trying to downplay

his own actions. A senior military officer at the Pentagon told *The Sunday Telegraph* on July 26, 2009:

> US policy is to fight these attacks as strongly as possible. As a result of Mr. McKinnon's actions, we suffered serious damage. This was not some harmless incident. He did very serious and deliberate damage to military and NASA computers and left silly and anti-America messages. All the evidence was that someone was staging a very serious attack on US computer systems.

McKinnon was first interviewed by the British police on March 19, 2002 and his computer was seized by the authorities. He was interviewed again on August 8 of that year, this time by the UK National Hi-Tech Crime Unit (NHTCU). Then in November 2002, McKinnon was indicted by a federal grand jury in the Eastern District of Virginia. The indictment contained seven counts of computer-related crime, each of which carried a potential ten-year jail sentence. However, McKinnon remained at large for the next three years and frequently talked to the media.

McKinnon told the British media that he obtained unauthorized access to computer systems in the United States including those mentioned in the United States indictment. He told the press that his motivation was drawn from a statement made before the Washington Press Club on May 9, 2001 by "The Disclosure Project" about government suppression of anti-gravity technology. McKinnon said he hacked into the Pentagon's computers to find evidence of UFOs, antigravity technology, and the suppression of "free energy," something that readers of my books probably know all about. McKinnon has stated that all of these things have been behind his actions.

In an interview televised on the BBC's *Click* program on May 5, 2006, McKinnon stated that he was able to get into the military's networks simply by using a Perl script that searched for blank passwords; in other words his report suggests that there were computers on these networks with the default passwords active.

In the 2006 interview with the BBC, he also said of "The Disclosure Project" that "they are some very credible, relied-upon

An alleged photo of part of the Solar Warden fleet circulating on the Internet.

people, all saying yes, there is UFO technology, there's anti-gravity, there's free energy, and it's extraterrestrial in origin and [they've] captured spacecraft and reverse engineered it."

On the BBC program McKinnon said he investigated a NASA photographic expert's claim that at the Johnson Space Center's Building 8, images were regularly cleaned of evidence of UFO craft. He confirmed this, comparing the raw originals with the "processed" images. He claimed to have viewed a detailed image of "something not man-made" and "cigar shaped" floating above the northern hemisphere. Assuming his viewing would be undisrupted owing to the hour, he did not think of capturing the image. He said he was "bedazzled," and therefore did not think of securing it with the screen capture function in the software, but then his connection was interrupted. This photo may be the one that can be found on the Internet and is reproduced here. That photo shows five large cylindrical spacecraft and three smaller craft in orbit together above Earth. This is supposedly a blurry photo a portion of the Solar Warden Space Fleet.

McKinnon continually told the media about the ships or craft within Solar Warden—the eight cigar-shaped Motherships (each longer than two football fields end-to-end) and 43 small "scout ships." He told the media that he stumbled upon a secret list of one hundred off planet US Naval officers and up to ten space warships after breaking into the computer databases. Interestingly, a couple of the spaceships were named USSS Curtis Lemay and USSS Roscoe

Hillenkoetter. Hillenkoetter was MJ-1 and the first head of the CIA..

McKinnon never really said that he found information about aliens or extraterrestrial contacts. The information that he found was basically about the secret space program Solar Warden. We know that McKinnon is sincere in what he has told the media and he also told the BBC and others that he was afraid that he would end up in Guantanamo Bay if he was to ever stand trial in a US military court.

McKinnon remained at liberty without restriction for three years until June 2005 (when the UK enacted the Extradition Act 2003, which implemented the 2003 extradition treaty with the United States wherein the United States did not need to provide contestable evidence). At that time he became subject to bail conditions, including a requirement to sign in at his local police station every evening and to remain at his home address at night.

If extradited to the US, McKinnon would have faced up to 70 years in jail. When he expressed his fears that he could be sent to Guantanamo Bay on British media he became a celebrity to the British public who felt he was just an innocent young computer genius with Asperger's syndrome who lived with his mom, and not a security threat.

According to Wikipedia, McKinnon's lawyers in the House of

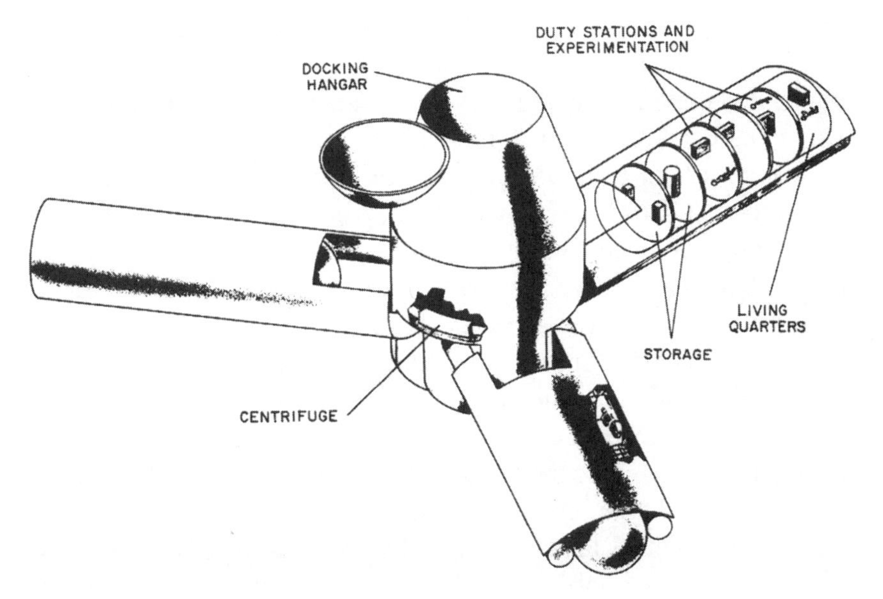

A 1963 NASA drawing of an orbital space station.

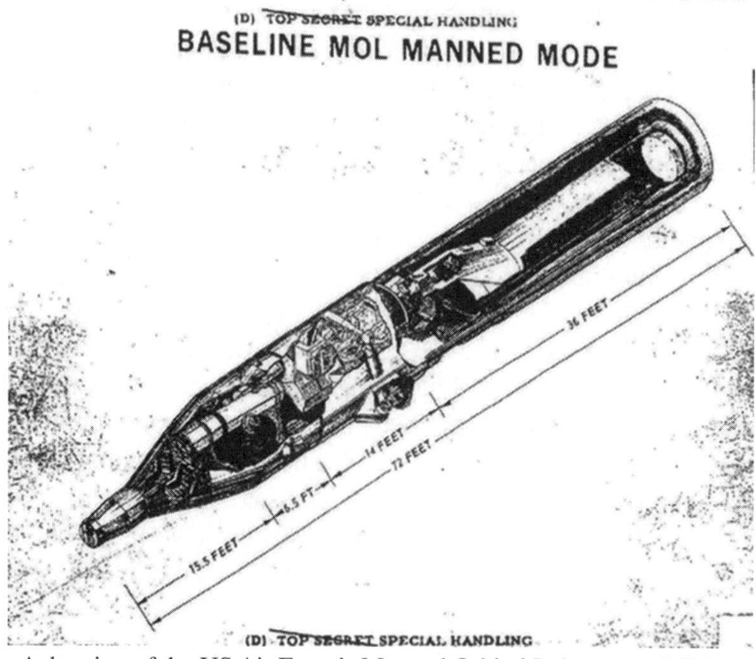

A drawing of the US Air Force's Manned Orbital Laboratory (MOL).

Lords on June 16, 2008, told the Law Lords that the US prosecutors had said McKinnon faced a possible 8–10 years in jail per count if he contested the charges (there were seven counts) without any chance of repatriation, but only 37–46 months if he cooperated and went voluntarily to the United States. McKinnon's lawyers contended that in effect this was intimidation to force McKinnon to waive his legal rights. McKinnon also stated that he had been told that he could serve part of his sentence in the UK if he cooperated. McKinnon and his lawyers rejected the offer because the Americans would not guarantee these concessions.

McKinnon's lawyer said that the Law Lords could deny extradition if there was an abuse of process: "If the United States wish to use the processes of English courts to secure the extradition of an alleged offender, then they must play by our rules."

However, the House of Lords rejected this argument, and then McKinnon appealed to the European Court of Human Rights, which briefly imposed a bar on the extradition, but the request for an appeal was rejected. On 23 January 2009, McKinnon won permission from the High Court to apply for a judicial review against his extradition.

Upon losing this appeal, McKinnon's legal team applied for a judicial review into the Home Secretary's rejection of medical evidence, which stated that, when he could easily be tried in the UK, it was unnecessary, cruel and inhumane to inflict the further stress of removing him from his homeland, his family and his medical support network. Finally, on October 16, 2012, then-Home Secretary Theresa May announced to the House of Commons that the extradition had been blocked, saying:

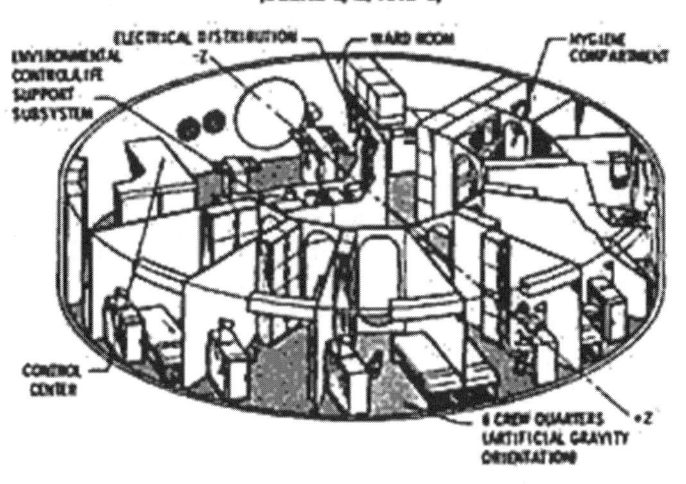

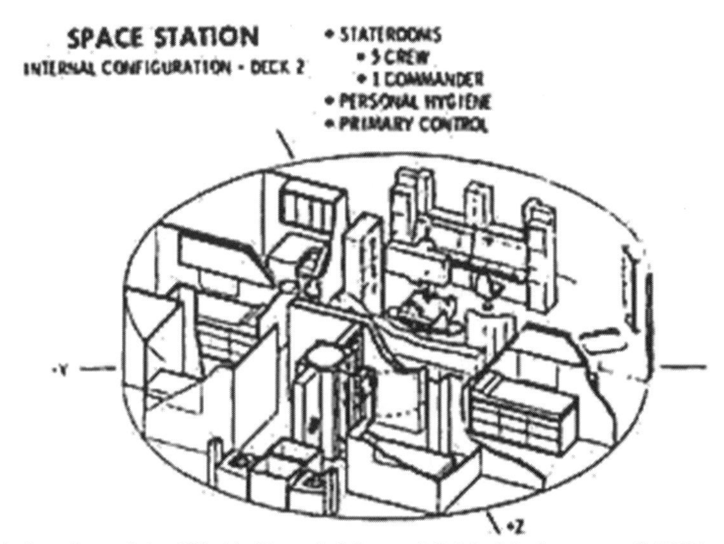

A drawing of the US Air Force's Manned Orbital Laboratory (MOL). This project was begun in 1963 and was cancelled in 1969.

A plastic model of the US Air Force's Manned Orbital Laboratory (MOL).

> Mr. McKinnon is accused of serious crimes. But there is also no doubt that he is seriously ill [...] He has Asperger's syndrome, and suffers from depressive illness. Mr. McKinnon's extradition would give rise to such a high risk of him ending his life that a decision to extradite would be incompatible with Mr. McKinnon's human rights.

She stated that the Director of Public Prosecutions would determine whether McKinnon should face trial before a British court. On December 14, 2012 it was announced that McKinnon would not be prosecuted in the United Kingdom because of the difficulties involved in bringing a case against him when the evidence was in the United States.

During all these years McKinnon gathered much support from the public and a number of celebrities. In November of 2008, the rock group Marillion announced that it was ready to participate in a benefit concert in support of McKinnon's struggle to avoid extradition to the United States. Many prominent individuals voiced support, including Sting, Trudie Styler, Julie Christie, David Gilmour, Graham Nash, Peter Gabriel, The Proclaimers, Bob Geldof, Chrissie Hynde, David Cameron, Boris Johnson, Stephen Fry, and Terry Waite. All proposed that, at the very least, he should be tried in the UK.

In August 2009, the Glasgow newspaper *The Herald* reported

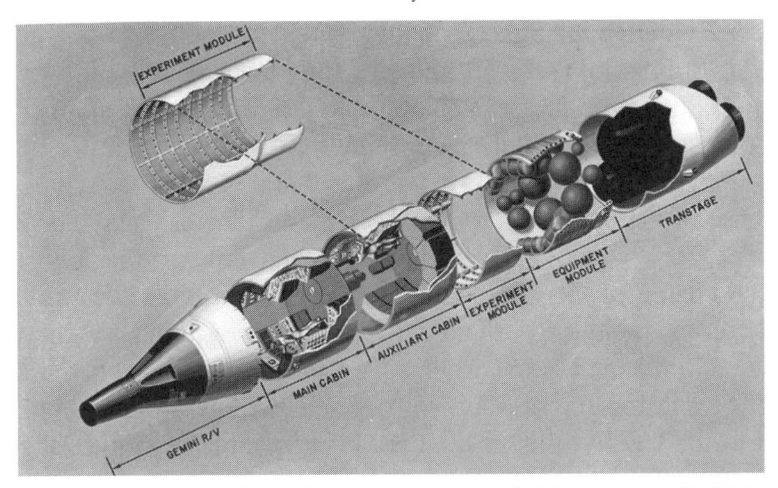

A drawing of the US Air Force's Manned Orbital Laboratory (MOL).

that the Scottish entrepreneur Luke Heron would pay £100,000 towards McKinnon's legal costs in the event he was extradited to the US. Web and print media across the UK were critical of the extradition, and *The Daily Mail* ran a campaign to prevent it.

Also in August of 2009, Pink Floyd's David Gilmour released an online single, "Chicago—Change the World," on which he sang and played guitar, bass and keyboards, to promote awareness of McKinnon's plight. A re-titled cover of the Graham Nash song "Chicago," it featured Chrissie Hynde and Bob Geldof, plus McKinnon himself. It was produced by long-time Pink Floyd collaborator Chris Thomas and was made with Nash's support.

According to Wikipedia, on July 20, 2010, Tom Bradby, the British broadcaster ITN's political editor, raised the Gary McKinnon issue with President Barack Obama and Prime Minister David Cameron at a joint White House press conference in Washington DC. Obama and Cameron responded that they had discussed McKinnon's situation and were working to find an "appropriate solution."

So, after a decade of international intrigue, the "biggest military computer hack of all time" could finally rest easy at his mom's house and not have to think about detention in Guantanamo Bay for the rest of his life. But what of Solar Warden? Do we have other evidence that it exists?

Other Evidence for Solar Warden

According to the well-known UFO investigator and writer George Filer, who has written about Solar Warden in a number of his blogs, Navy hospitals have numerous patients claiming they were part of Solar Warden. He believes that it is real. Filer says that because of her support for Gary McKinnon, Theresa May became the new Conservative Party leader and second female prime minister, taking charge of the UK on July 24, 2016. This was because she had become famous in Britain when the Obama Administration was trying to force the UK to extradite McKinnon to face trial in the US. Filer says it was Teresa May who refused to allow McKinnon to be sent to the US and this is what made her popular with the British voting public.

Filer also mentioned that there is an interesting document concerning President Reagan that came to light when the National Archive Records Administration made available 250,000 pages of documents from President Reagan's administration. Filer says that the entry for June 11, 1985 (page 334) reads:

> Lunch with 5 top space scientists.
>
> "It was fascinating. Space truly is the last frontier and some of the developments there in astronomy etc. are like science fiction, except they are real. I learned that our shuttle capacity is such that we could orbit 300 people."

Filer says that this would indicate a space ship at least as large as a 747-8 aircraft that is 250 feet long. It is likely much bigger to accommodate sleeping quarters, kitchen facilities, bathrooms, storage, radar and weapons. The now grounded Space Shuttle held a maximum of eight people and only five were built for space flight.

Filer says that apparently President Reagan revealed the existence of a highly classified space program that could accommodate hundreds of astronauts in orbit. This space program is most probably Solar Warden.

Filer then says that this Navy space fleet could be used to essentially clean up all the space debris that is floating out there in orbit around the Earth, causing potential danger to space stations and satellites, even large debris that NASA or others would normally let

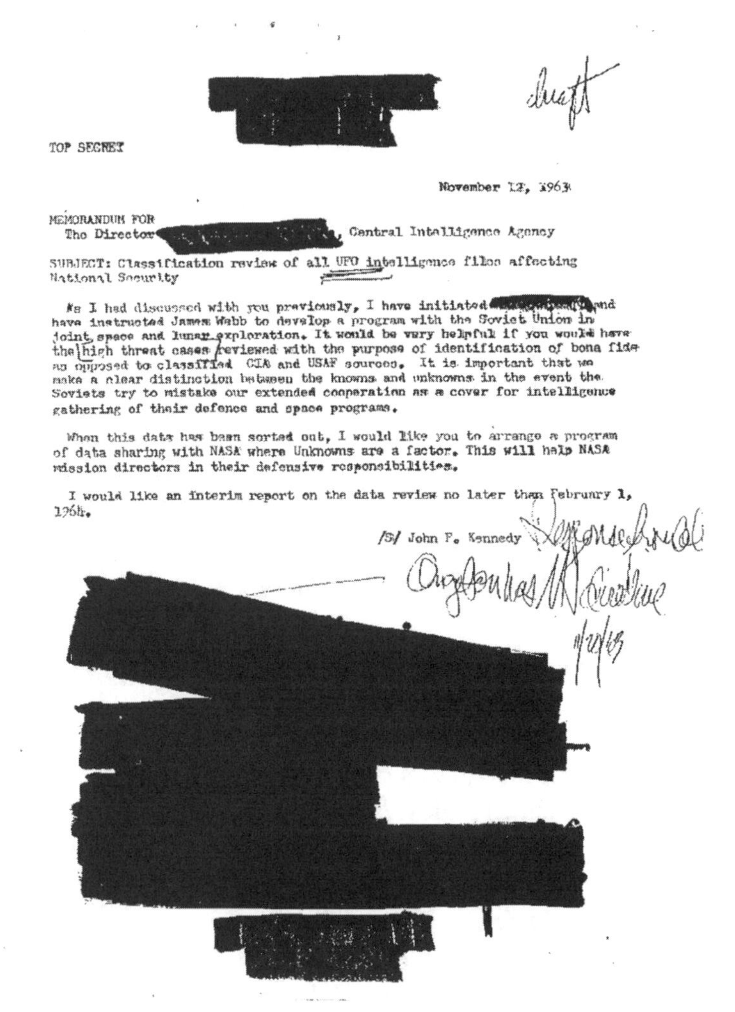

TOP SECRET

November 12, 1963

MEMORANDUM FOR
The Director, Central Intelligence Agency

SUBJECT: Classification review of all UFO intelligence files affecting National Security

As I had discussed with you previously, I have initiated and have instructed James Webb to develop a program with the Soviet Union in joint space and lunar exploration. It would be very helpful if you would have the high threat cases reviewed with the purpose of identification of bona fide as opposed to classified CIA and USAF sources. It is important that we make a clear distinction between the knowns and unknowns in the event the Soviets try to mistake our extended cooperation as a cover for intelligence gathering of their defence and space programs.

When this data has been sorted out, I would like you to arrange a program of data sharing with NASA where Unknowns are a factor. This will help NASA mission directors in their defensive responsibilities.

I would like an interim report on the data review no later than February 1, 1964.

/S/ John F. Kennedy

A 1963 letter from President Kennedy to the CIA about UFO activity.

fall eventually to Earth. All could be blasted into tiny particles with powerful lasers by the Navy's space force. Says Filer:

> Normally, NASA vehicles would require gradual orbital corrections that would take much time and be insufficient to deal with an immediate threat. According to Ted Twietmeyer, the citation is circumstantial evidence for the existence of antigravity vehicles with advanced particle beam weapons that could remove orbital debris from the path of NASA vehicles. These space ships must be armed with particle

> beam or quantum weaponry, chemical lasers or electrically excited reactor powered lasers and other weaponry. These weapons could also be used to destroy objects in the path of the space station or space shuttle. There were several eyewitness reports of hovering black triangles firing at targets in Iraq during the early days of the war.

So, here we have some corroboration that a space force has existed for many decades. Does this space force have several bases on the Moon? It is not inconceivable that this space force has been to Mars and possibly other planets.

Some people have apparently sought to capitalize on the revelations of super hacker Gary McKinnon about Solar Warden. One such person is Randy Cramer who claims that he was a captain in the Marines and was sent to Mars in 1987 where he spent the next 17 years helping defend bases on that planet from "reptoid and insectoid aliens."

Cramer claims in the many interviews that he has done, that he was trained as a child super soldier to serve as a member of an elite Marine Corps unit that provides personnel for a secret space program with military bases on the Moon, Mars and in other parts of the solar system. In these interviews Cramer describes how he traveled to a secret a Moon base to sign papers committing him to a 20-year tour of duty. He claims that he was part of secret space force called the Earth Defense Force (EDF).

He also claims that there is a Mars Defense Force (MDF) that was created by the Mars Colony Corporation that has five civilian settlements on Mars. The MDF was to protect these five human settlements from indigenous Martians, who are reptoids and insectoids. Cramer claims he flew to Forward Station Zebra, a base on Mars, and spent the next 17-plus years of his life as a member of the special tactical operations division.

The Mars Colony Corporation's mission is to extract minerals; it uses slave labor—people who are kidnapped from Earth. Cramer says that his job was to defend the operation from Martians. He says his team fought a Martian reptoid species and a Martian insectoid species while on Mars and these two species have their own nests and hives. Cramer says that there was a third species, also reptoids, called the Draconians, who were invaders from the star system of Alpha Draconian. Cramer maintains that the MDF eventually made treaties with the native Martian reptoids and insectoids, and they formed a coalition to fight the Draconians.

After 17 years on Mars, Cramer says that he returned to a base on the Moon where he was age-regressed back his age in 1987 and reinserted into the human population on Earth. So, nobody even knew he had been on Mars for 17 years, and he resumed a seemingly normal life.

Cramer's story is written about in Len Kasten's 2020 book *Dark Fleet*,[64] and appears in other books as well. His story sounds like part of the plot for the 1990 movie *Total Recall* starring Arnold Swartzenegger. Kasten received a letter from Alfred Lambremont Webre who said:

> I have been informed that on November 6, 2019, NASA Astronaut Ken Johnston Sr., a US Marine Corps veteran, officially inquired of the responsible US Marine Corps Command Staff Sgt Nelson and Staff Sgt Cartwright at Los Lunas, N.M. Marine Base https://loslunasnm.gov/ and was duly informed that "there is no record of a person named Randy Cramer ever having served in the U.S. Marine Corps."
>
> These USMC personnel first searched for any US Marine named Randy Cramer having served in the USMC between 1987 and 2004. These were the dates of the 17 years that Randy Cramer claims in your book and publicly for the last seven years since 2014 to have served on Mars. These USMC personnel also searched for whether the USMC has ever had a Captain rank by the name of Randy Cramer in its entire history. This search also came up Negative, when the entire database of USMC personnel was searched.
>
> Assuming that this prima facie evidence is in fact true, that would mean that at the very least Randy Cramer could be found guilty of a federal felony and violation of the **Stolen Valor Act** (https://www.congress.gov/bill/113th-congress/house-bill/258/) for publicly putting himself forth as a military officer, not to mention the untold damage to

exopolitical research of perpetrating a sustained public hoax and fraud about his alleged participation in the U.S. presence on Mars.

...I understand as well that Andrew D. Basiago, Esq., a member of the Washington State Bar and the US District Court for the Western District of Washington has found a serious distortion of his Mars experiences in every paragraph of the article you wrote about him for *Atlantis Rising*.

In addition, Andrew D. Basiago has spoken with the former wife of Randy Cramer with regard to Cramer's alleged service in the US Marine Corps and on Mars. Randy Cramer's former wife reportedly told Andrew D. Basiago that "Randy Cramer has never been in the Marine Corps, he has never been on Mars. He is a former bartender who has made up these fictions for profit and attention, and because he is a pathological liar, which was the cause of marital breakdown." For full details, I urge you to call U.S. Chrononaut and Mars Astronaut Andrew D. Basiago. Needless to say, I would urge you to delete any and all references to Randy Cramer in the final pre-publication draft of your book, *Dark Fleet*.

Please let me know if you need any information.

Best regards,

Alfred Lambremont Webre, JD

So, according to Cramer's former wife, he is really a bartender and habitual liar who was never in the Marine Corps at all. His story is so unverifiable that it must be considered a hoax. He has 17 missing years on Mars that are not missing at all and his story is largely unbelievable, including the battles on Mars that he says he took part in. It is clearly a mishmash of the popular stories about reptilian extraterrestrials and a secret space program. The people who are endorsing him are the same people who always endorse these stories and often make similar claims, such as being secretly trained as a space warrior, starting as a child, or being invited for some reason to go to Mars and participate in the secret space program—but never actually going. Sadly, people do make up stories to get attention or for other reasons.

However, the tales of Gary McKinnon are completely different. McKinnon did not seek the limelight, rather it was thrust upon him and he and his family were genuinely in fear of his spending the rest of his life in prison.

Yes, there is a secret space program—probably several. One apparently is code named Solar Warden. Just what is going on with these secret space programs? What are these extra-terrestrial officers doing up in space? Where are the launch and other staging areas for these hundreds of personnel who are going into space? It doesn't appear that these launches are happening at Area 51 in Nevada or at any location in the United States. Is it possible that Antarctica is the staging area for Solar Warden and other secret space programs?

(19) **United States**
(12) **Patent Application Publication** (10) **Pub. No.: US 2019/0058105 A1**
Pais (43) **Pub. Date: Feb. 21, 2019**

(54) **PIEZOELECTRICITY-INDUCED ROOM TEMPERATURE SUPERCONDUCTOR**

(71) Applicant: **Salvatore Cezar Pais**, Callaway, MD (US)

(72) Inventor: **Salvatore Cezar Pais**, Callaway, MD (US)

(73) Assignee: **United States of America as represented by the Secretary of the Navy**, Patuxent River, MD (US)

(21) Appl. No.: **15/678,672**

(22) Filed: **Aug. 16, 2017**

Publication Classification

(51) **Int. Cl.**
H01L 41/107 (2006.01)
H01L 41/08 (2006.01)
H01B 12/06 (2006.01)

(52) **U.S. Cl.**
CPC ***H01L 41/107*** (2013.01); ***H01B 12/06*** (2013.01); ***H01L 41/0805*** (2013.01)

(57) **ABSTRACT**

The present invention is a room temperature superconductor comprising of a wire, which comprises of an insulator core and a metal coating. The metal coating is disposed around the insulator core, and the metal is coating deposited on the core. When a pulsed current is passed through the wire, while the wire is vibrated, room temperature superconductivity is induced.

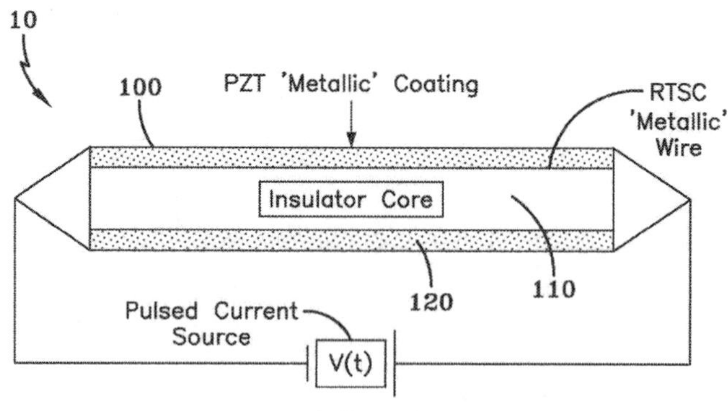

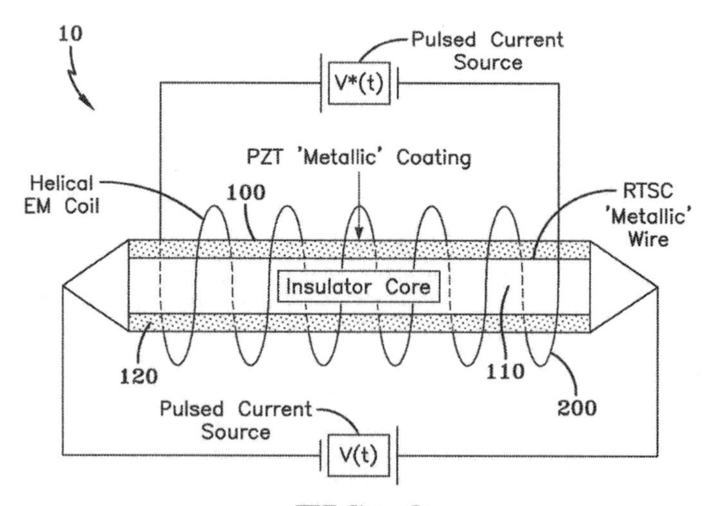

A US Navy patent for a room temperature superconductor from 2019.

Chapter Nine

Antarctica and the Secret Space Program

In the dark of the sun will you save me a place?
Give me hope, give me comfort, get me to a better place?
I saw you sail across a river underneath Orion's sword
In your eyes there was a freedom I had never known before.
Hey, yeah, yeah, in the dark of the sun
We will stand together
Yeah we will stand as one in the dark of the sun.
—*The Dark of the Sun*, Tom Petty

We have looked at the strange activities in Antarctica during and after WWII, as well as the US military's plans to put secret bases on the Moon. We have covered the tale of Gary McKinnon and Solar Warden. Does any of this have to do with Antarctica? Let us take a look at this frozen continent in the decades since WWII and the strange goings on in the 1950s.

After Operation Highjump (1946–1947) there was Operation Windmill (1947–1948). The US Navy did not return to Antarctica for seven years but in 1955 it began Operation Deep Freeze which would last until 1956. However, Operation Deep Freeze was to be an ongoing operation lasting decades, with different ships and personnel rotating in their duties and patrols in the waters around Antarctica as well as on the continent itself. The Operation Deep Freeze activities were to be succeeded by "Operation Deep Freeze II," and so on.

The impetus behind Operation Deep Freeze I was the International Geophysical Year 1957–58. The International Geophysical Year was a collaborative effort among forty nations to carry out earth science studies from the North Pole to the South Pole and at points in between. The United States, along with New Zealand, the United Kingdom, France, Japan, Norway, Chile, Argentina, and the USSR agreed to go to the South Pole, the least explored land on Earth.

Their goal they said was to advance world knowledge of Antarctic hydrography and weather systems, glacial movements, and marine life. Operation Deep Freeze I prepared a permanent research station and paved the way for more exhaustive research in later Deep Freeze operations. The expedition consisted of seven ships, including three icebreakers, and was known as Task Force 43.

On October 31, 1956, US Navy Rear Admiral George Dufek and others successfully landed an R4D Skytrain (Douglas DC-3) aircraft at the South Pole, as part of the many expeditions mounted for the International Geophysical Year. This was the first aircraft to land at the South Pole and the first time that Americans had set foot on the South Pole. This marked the beginning of the establishment of the first permanent base, by airlift, at the South Pole. The base was commissioned on January 1, 1957 and named "Old Byrd." Today this base is known as the Amundsen–Scott South Pole Station.

An astounding occurrence was seen by staff during the extended

UFO SMASHES THROUGH POLAR ICE

MANY NAVAL experts are convinced that UFOs operate from bases deep beneath the ocean.

The theory received a dramatic boost during the US Navy's Operation Deep Freeze in the Antarctic.

While crew-members aboard an icebreaker in Admiralty Bay watched amazed, an immense silvery craft smashed upward through thick ice and hurtled into the night sky.

According to the ship's log, the ice through which the bullet-shaped object sliced was at least 12m thick.

Huge blocks of frozen seawater, hurled high into the air, came cascading down around the exit hole, in which the ocean boiled and spouted steam.

This encounter was only one of hundreds reported between ships and UFOs in the past 40 years.

So many saucers have appeared above seas around Australia that in 1966 the US Navy sent a Professor McDonald here to study them. His findings, about "floating reefs" and glowing globes of light seen plunging into Bass Strait, have never been publicly released.

But the sheer weight of evidence suggests that something very strange is going on beneath Australia's oceans. Many sea captains and their officers have reported seeing gigantic, seemingly alien craft near their ships.

- A typical witness was Captain Julian Ardanza of the Argentine ship Naviero. While sailing off the Brazilian coast, he and his crew saw a "huge, glowing cigar" on the starboard side.

"It made no noise and there was no sign of periscope.

Operation Deep Freeze in Antarctica on March 16, 1961 and reported in newspapers and magazines from 1961 to 1966. Newspapers around the world carried the story of a silvery, bullet-shaped UFO crashing up out of the ice near the coast of Antarctica, scattering tons of ice into the air and flying off into the sky. The brief but sensational story is represented by this story from an Australian magazine in 1966, which starts with the headline, "UFO Smashes Through Polar Ice":

> Many naval experts are convinced that UFOs operate from bases deep beneath the ocean.
>
> The theory received a dramatic boost during the US Navy's Operation Deep Freeze in the Antarctic. While crewmembers aboard an icebreaker in Admiralty Bay watched amazed, an immense silvery craft smashed upward through thick ice and hurtled into the night sky.
>
> According to the ship's log, the ice through which the bullet-shaped object sliced was at least 12 meters thick.
>
> Huge blocks of frozen seawater, hurled high into the air, came cascading down around the exit hole, in which the ocean boiled and spouted steam.
>
> The encounter was only one of hundreds reported between ships and UFOs in the past 40 years.
>
> So many saucers have appeared above seas around Australia that in 1966 the US Navy sent a Professor McDonald here to study them. His findings about "floating reefs" and glowing globes of light seen plunging into Bass Strait, have never been publically released.
>
> But the sheer weight of evidence suggest that something very strange is going on beneath Australia's oceans. Many sea captains and their officers have reported seeing gigantic, seeming alien craft near their ships.
>
> A typical witness was Captain Julian Ardanza of the Argentine ship *Naviero*. While sailing off the Brazilian coast, he and his crew saw a "huge glowing cigar" on the starboard side.

This Australian article contains the curious mention of the

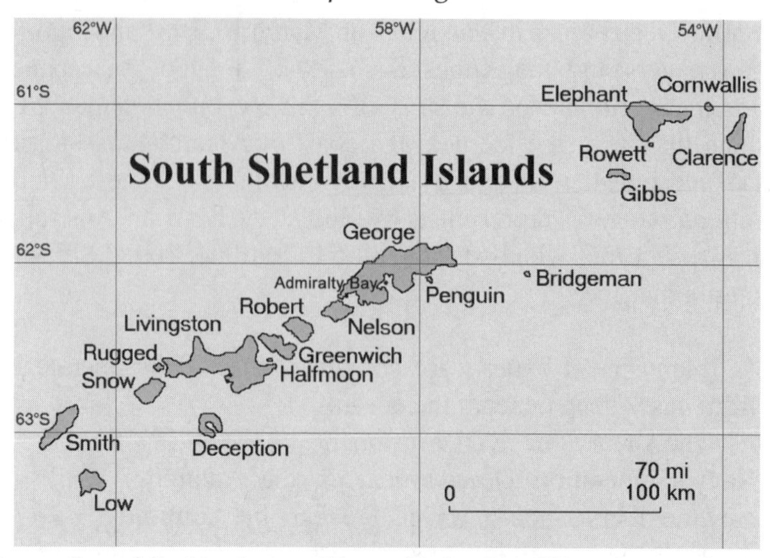

Argentine ship *Naviero* sailing off the Brazilian coast where the crew saw a "huge glowing cigar." This seems to be one of the "submarines that can fly" that were revealed in earlier chapters. This may be the Andromeda craft shown in diagrams and photos in the 1989 document dump.

This same story about Operation Deep Freeze is told by Ivan T. Sanderson in his book about underwater UFOs (USOs), *Invisible Residents*.[57] Sanderson quotes from the March 1966 issue of *Man's Illustrated* which had an article entitled "U.F.O.s—at 450 Fathoms!" by Ed Hyde. The exact day of the sighting is never given, but it apparently occurred in the evening on March 16, 1961. The witness was a Brazilian scientist named Dr. Rubens J. Villela who was on an icebreaker during the ongoing Operation Deep Freeze. Sanderson says that Dr. Villela was in the command cabin on watch when he was "literally jolted almost out of both his body and his mind by a "something" that suddenly came roaring up out of the sea through no less than 37 feet of ice, and went on up into the sky like a vast silvery bullet." The incident occurred at Admiralty Bay which is an irregular bay on the southern coast of King George Island (just called George Island on many maps) in the South Shetland Islands of Antarctica. The Hyde article quoted by Sanderson states:

> The ship was in the Admiralty Bay, which faces the South Atlantic Ocean. The only other witnesses were the

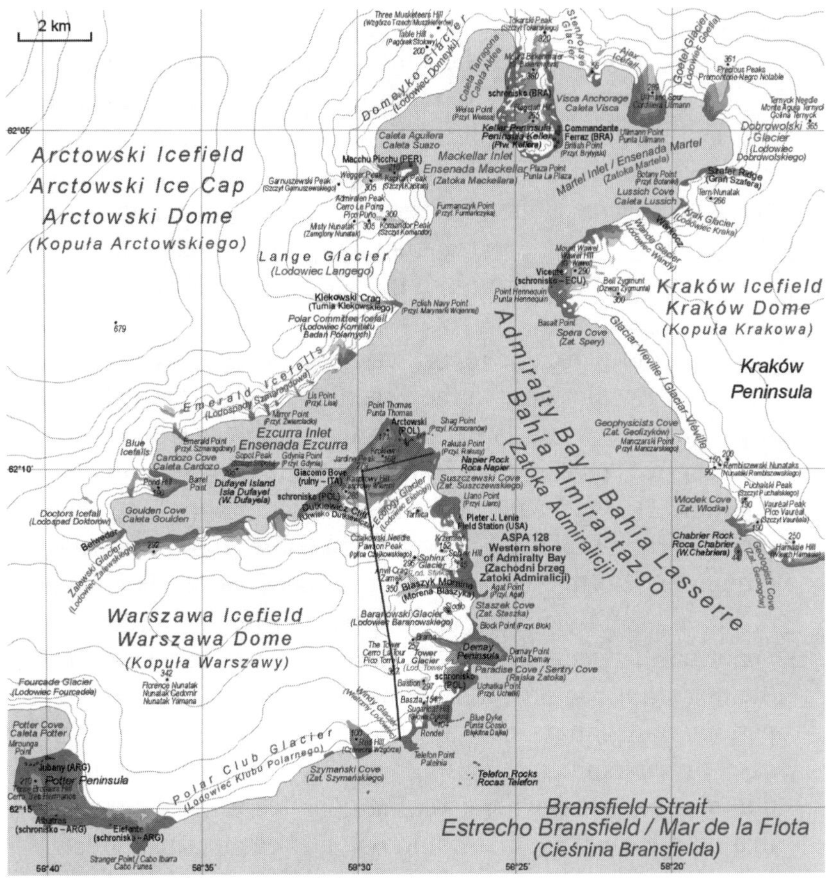

A map of Admialty Bay on King George Island in the South Shetland Islands.

officer on watch and the helmsman, as it was an extremely cold day and all other personnel were below decks. Further, these witnesses saw only the tag end of the performance because they were busy with charts; but what they did witness was quite enough. Enormous blocks of ice had been hurled high into the air and came cascading down all around the hole burst through the thick ice-sheet, and the water was rolling, and apparently boiling, while masses of steam issued from both the hole and descending ice. From Dr. Villela's description, the spectators were apparently "not amused." Nor are we.

Continues Sanderson:

> First of all, this is the kind of report that aggravates everybody, and notably reporters—and for sundry reasons that I will take up in a moment. Second, it infuriates scientists and technologists, and possibly even the Navy, since the story was never officially confirmed. Finally, it unnerves the general public, and particularly the better-informed and those of enquiring mind. In fact, it is an all-round and damned nuisance to everybody.
>
> To begin with, this report was issued quite formally by the US Navy, but only to the Brazilian press as far as I can ascertain. When the latter published it, and the European wire services picked it up, it was promptly denied in official quarters. However, the Spanish-speaking Latin-American newspapers reproduced it, notably in Chile, and from these airings the North American wire services ran it merely as a "filler"—and only once. Then, everything returned to abnormal. To get at even these facts has taken us years, and the net result has proved to be remarkably bipartisan. One party simply shrugs and says "cover-up"; the other, in this case officialdom, says just as simply "rubbish." Neither attitude gets us anywhere; yet I not only contend, but insist, that reports such as this should not, and cannot, just be left lying around in newspaper morgues. The validity of the facts stated constitutes a fact in itself, and in this case it cries out for explanation.[57]

So the story that we might infer from Sanderson is that because Dr. Villela was a Brazilian scientist and not part of the US military, he was free to speak about the astonishing event he had witnessed. So, when the ship returned from Antarctica and docked in Brazil to let Dr. Villela and other Brazilians off the ship, the scientist went to the Brazilian media to tell his story and when the fascinated Brazilian media asked the US Navy officers on the ship whether the report was true or not, they did not deny.

Within a few weeks and months of the story being reported in Chile, Argentina and Europe, interest in the story increased and apparently some investigators—perhaps Sanderson himself, as he was former British Navy and British Intelligence, and a close friend of James Bond creator Ian Fleming—asked the US Navy if there was an official report on the incident within Operation Deep Freeze and they denied that there was. So, officially the US Navy is denying the report, but the story was widely reported around the world in 1966. We seem to have to rely on the early newspaper stories in Brazil and Chile for the scant information that we have. It is worth noting that both countries—and all countries in South America, for that matter—are keenly interested in UFO reports, and South American media is constantly reporting on UFO sightings and other anomalous activity. The reality of UFOs is not so widely questioned in South America—not to say Antarctica—as it is in the USA, Britain or Canada.

I was able to do some digging on the Internet and I found that Dr. Villela has written a number of articles on his sighting, and on other UFO incidents in Brazil and the southern Atlantic starting with a 1961 article entitled: "Operação Congelada: relato de uma viagem ao continente gelado" (Operation Deep Freeze—report on a voyage

A 1970s drawing of the incident at Admiralty Bay, although the craft was described more like a bullet or tubular craft than a pointed missile like this.

to the frozen continent), which appeared in *Folha de S. Paulo*, 20 May to 11 June (a 20 part series). Then in 1968 he wrote an article on the incident for the French magazine *Phénomènes Spatiaux* where the title gives the date of his famous sighting: "Baie de l'Amirauté, 16 mars 1961: une observation par M. Rubens Junqueira Villela," *Phénomènes Spatiaux* 16: 17–23.

Then in 1979 he wrote an article in Portuguese for the Brazilian magazine *Disco Voador* (*Flying Saucer*) that was entitled: "Envolvimento crescente em quatro observações de OVNIs" (Increasing involvement in four UFO observations). He seems to have spent quite a bit of time in the Antarctic and published a scientific article in 1991 entitled: "Radio weather transmissions in the Antarctic," in the *Polar Record 27* (161): pages 103–114.

Finally, he wrote an article for the Brazilian magazine *Revista UFO Brasil* in 1998 entitled "Discos voadores na Antártida" (Flying saucers in Antarctica), *Revista UFO Brasil* 58: 22–27 (May). This article was reproduced in English as "UFOs in Antarctica" in *UFO Magazine* (UK), November–December 1998: pages 10–13.

By looking at the location of incident, we might learn a little more about what was going on here. Here's what Wikipedia as to say about Admiralty Bay:

> …The name appears on a map of 1822 by Captain George Powell, a British sailor, and is now established in international usage. The Henryk Arctowski Polish Antarctic Station is situated on the bay, as is the Comandante Ferraz Brazilian Antarctic Base. It has been designated an Antarctic Specially Managed Area.
>
> The bay has three fjords: Martel, Mackellar, and Ezcurra. A mariner's guide to the region pronounced the bay to have the best anchorage of any in the South Shetlands, "being well-sheltered all around and having moderate depths over a bottom of good, stiff clay. Ice from the glaciers is frequently troublesome."

So, we might have found something here. We know that the Germans preferred the steep granite walls of fjords in Norway for their U-boat bases. The secret U-boat base in the Canary Islands was where cliffs met the ocean. Here on King George Island was the same sort of geography—deep water with steep cliffs where secret submarine pens can be created inside of a coastal mountain. This is exactly the sort of place that many countries seek for their secret navies and hidden bases.

The Strange Case of Carl Disch

Carl Robert Disch was a German American scientist who disappeared in Antarctica in 1966. He is something of a mysterious character, aside from his disappearance; his birth date is unknown. There is a memorial for him at the Greenwood Cemetery in Monroe, Wisconsin. Disch vanished from his substation in Antarctica on May 8, 1965.

Disch worked as an ionospheric physicist at Byrd Station, Antarctica. He was on a team that was investigating radio noises for the National Bureau of Standards. He was working at the radio noise building, which was situated about 7,000 feet, just over a mile from the main station complex.

At 09:15 on the morning of May 8 Disch left the building, and set off with a purposeful tread to the main complex. It was a journey he had done 25 times before, and so was very familiar with it. The temperature at that time was a forbidding -45 degrees F (-42 degrees C), but Disch was well equipped in his Polar gear. When he hadn't arrived at the main complex by 10:00 AM, a vehicle search party was organized. At 11:30 his trail was picked up leading to the southwest corner of the skiway, about four miles away. The search party returned to base to refuel, and then spent the next three hours trying to pick up the trail again. He was never found.

According to the government website. www.boulder.nist.gov, this is what happened to Carl Disch:

> Carl R. Disch, ionospheric physicist for the National Bureau of Standards Boulder Labs, disappeared on May 8, 1965, from the Ionosphere-Forward

Carl Disch.

Carl Disch with a beard in Antarctica.

Scatter station near Byrd station in the Antarctic. Disch was a member of the Boulder Laboratories' 1964-65 Antarctic research team. The team was spending the year as part of NBS's contribution to the Year of the Quiet Sun (the period of low ebb for solar activity).

Disch was returning to the main station after a visit to the radio-noise installation when he apparently missed a handline. The temperature at the time was -45 F, with strong winds. In spite of the severe weather conditions the station personnel mounted a lengthy and thorough search for Carl. Not a trace was ever found.

"In all there were 27 (19 military and 8 scientists) and one dog who wintered over. Unfortunately, we lost

> a scientist by the name of Carl Disch and our Husky dog during the winter months. Carl wandered away from the 'life line' that connected the weather tower to the main tunnel on May 8th and Sastrugis "Gus" disappeared August 18. Far as I know they've not been found. We tried in vain to find Carl but at the time the wind was raging with minus 45 [degree] temperature. All we could do was tie ropes around our waists, spread out on both sides of the D-8 cat and walk along hoping to stumble over him. Was like finding a needle in a haystack blindfolded." —Jim Bartley, a former winter-over resident of Byrd Station. He spent 13 months there in 1964 and 1965)
>
> A memorial service for Carl Disch was held in his hometown of Monroe, Wisconsin on May 14, 1965, attended by Dr. T.O. Jones, Head of the Office of Polar Programs, and Stephen Barnes from the Central Radio Propagation Laboratory, National Bureau of Standards.

The main theories on this case are: a) He got lost in the weather conditions—but some people question this as he was quite experienced in his line of work. b) He fell into a crevasse somewhere. c) He intentionality left—suicide—but why though? d) Alien abduction—this is out there because his footprints just stopped and didn't continue. There is also a story of some lights—more on that below. e) He was working for or abducted by the Soviet Union—he was out in Antarctica doing his work during the Cold War. Some believe he was either abducted for what he knew or taken away by the Soviet Union because he was secretly working for them and his cover was about to be blown.

Some have claimed he got into an argument during a card game and walked out of the station in a flurry of anger, however, his departure was in the morning, not in the afternoon or evening when card games were typically played.

Perhaps he was actually working with the Third Power and was taken away by them. He was of German descent and spoke German.

The former California policeman David Paulides discusses the disappearance of Disch in his 2017 book *Missing 411: Off the Grid*,[62] one of a series of books about people who have mysteriously disappeared in Canada and the United States, plus a few disappearances in other countries. The disappearance of Carl Disch is his only case study of something that occurred on the polar continent.

Paulides quotes from John Keel's 1971 book *Our Haunted Planet*:

> The search went on for three days and covered a thirty-five-mile area around the hut. Disch's own dog, a Husky named Gus, disappeared shortly afterward. Some of the searchers claimed they saw mysterious lights and heard engine noises in the distance. Antarctica is of course uninhabited except for a handful of international scientists who work very closely with one another.

It should be noted here that the dog, Gus, disappeared three months later. Paulides was naturally curious about the lights that were supposedly seen during the search for Carl Disch and says he looked for more information on this aspect. He found more information in a November 24, 1966 article in the *Indianapolis Star* that had a confirmation of the rumors of lights in the sky:

> "It's a good Antarctic story," said Ron Sefton, the scientific leader at Byrd Station. "I've heard a number of versions myself, embellished with weird stories about the possibilities of the scientist and the dog being snatched up by UFOs, or the possibility of their heading for a hidden lush valley where folklore says the skuas (birds) go during the long Antarctic winter. But it didn't quite happen that way."
>
> … "If Disch had fallen and was lying on the snow, the husky would have seen him long before the searchers would. Similarly, if he had fallen and was covered by drifting snow, the dog would have sighted the mound and rushed out to investigate it. That's the way Huskies are."[62]

David Paulides, a trained investigator, thinks that Disch's disappearance is rather suspicious. He cites Disch's German ancestry as part of a pattern in disappearances in Canada and the

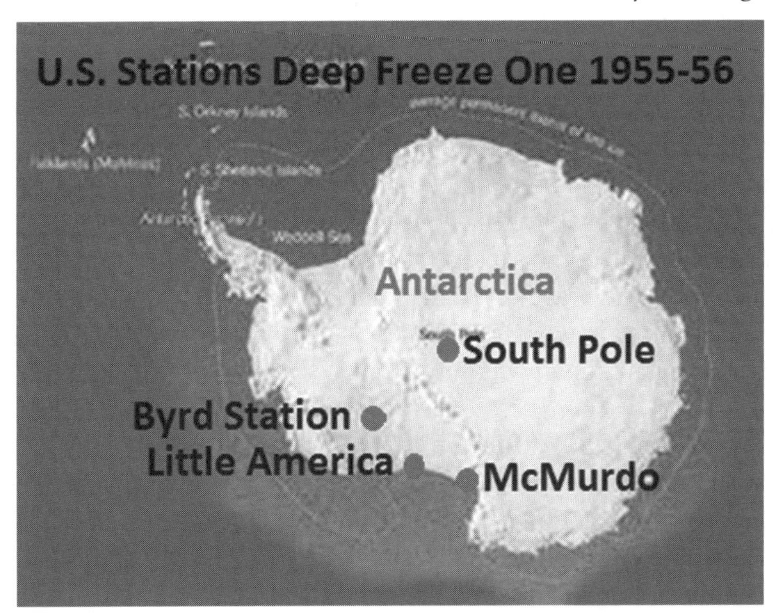

US, plus that he was a physicist and brilliant scientist, that his dog at first could not find the scent of Disch and then, three months later, vanished as well. As Paulides dryly points out, Disch has never been found.

Paulides also mentions the bizarre rumors from the Internet, unsubstantiated he says, that various research facilities in Antarctica have received radio communications from Disch in the years after his disappearance. Paulides could not find any credible sources for this claim, and it would be very strange indeed if Disch had somehow contacted other research facilities in the Antarctic after he vanished.

In 1971, it is claimed, a message arrived at McMurdo Message Center via the AA2 Weather Circuit. The author of the message claimed to be Carl Disch. The message said:

> To the world I am dead. They believe that my body is but a pinpoint frozen here to the surface of this white continent. I say to you, I, Carl Disch, live. Do not for one moment think that it was a mistake. Everything was planned. They pushed me, tormented me and bored me with their shallow lives. …the endless singing of the wind almost drives me mad. I begin to long for human companionship.

Many see this message as a hoax, one for tormented and bored Antarctic researchers. Still, one cannot help but think that Disch was taken from the base, perhaps by flying saucer, either willingly or unwillingly. Had he planned to disappear on that snowy day and then later come back for his dog? Had he been abducted against his will and taken to some secret Antarctic base? Was the reason he was abducted that he was a brilliant scientist, and of German descent? Or was he already working for these people and in touch with them throughout his time in Antarctica?

The Mystery of Byrd Station

Byrd Station, where Disch vanished from in May of 1965, is a former research station established by the United States during the International Geophysical Year in central Marie Byrd Land in 1956. Named in honor of American Antarctic explorer Admiral Richard E. Byrd, the station was closed in early 2005 and is currently an "abandoned" base, according to the US Military.

Says the website, http://www.westarctica.wiki/index.php/Byrd_Station:

> A joint Army, Navy, Air Force, and Marines operation supported an overland tractor train traverse that left out of Little America V in late 1956 to establish the station. The train was led by Army Major Merle Dawson and completed a traverse of 646 miles (1,040 km) over unexplored country in Marie Byrd Land to blaze a trail to a spot selected beforehand. The station consisted of a set of four prefabricated buildings and was erected in less than one month by US Navy Seabees.
>
> It was commissioned on 1 January 1957. The original station ("Old Byrd") lasted about four years before it began to collapse under the snow. Construction of a second underground station in a nearby location began in 1960, and it was used until 1972. The Operation Deep Freeze activities were succeeded by "Operation Deep Freeze II", and so on, continuing a constant US presence in Antarctica since that date. The Coast Guard participated, USCGC

Northwind supported the mission throughout the 1970s, 1971-72, 1972-73, 1976-77, 1979-80. The Navy's Antarctic Development Squadron Six had been flying scientific and military missions to Greenland and the arctic compound's Williams Field since 1975.

In early 1996, the United States National Guard announced that the 109th Airlift Wing at Schenectady County Airport in Scotia, New York was slated to assume that entire mission from the United States Navy in 1999. The 109th operated ski-equipped LC-130s had been flying National Science Foundation support missions to Antarctica since 1988. The Antarctic operation would be fully funded by the National Science Foundation. The 109th expected to add approximately 235 full-time personnel to support that operation. The station was then converted into a summer-only field camp until it was abandoned in 2004-05.

While most research stations in Antarctica are placed along the coastal regions, Byrd Station is located far inland, but not near any mountains or other geographic features that might be of value to science. George Toney, the scientific leader at Byrd Station in the 1957, speculated that the reason for the camp's inland location was that it was "spang in the middle of a huge unclaimed wedge of Antarctica where the

A tunnel going under the ice at Byrd Station.

United States might well launch a claim later on, if it came to that."

So, Byrd Station, essentially an under-ice base, has been abandoned since early 2005. Yet, it is unclear what Byrd Station was doing out there in the wastes of Antarctica in the first place. No one seems to really know why the joint US military base was put at that location, but it may have been to lay claim to a huge area of ice—away from any mountain ranges or other geological features.

What might have been other reasons for the joint staff at the Pentagon to want to place Byrd Station at this especially desolate location? Was it to have an under-ice base that could be used as a secret space communication base? Did the location have to do with some sort of world energy grid, placed at a point where the harmonic values and other esoteric-mathematical values would be auspicious? Was it near a wormhole or portal in time? In other words, was there something special about this little spot in the vast area of Antarctica? Perhaps the "specialness" of this spot is part of the reason that Carl Disch disappeared. Is this part of Antarctica some sort of "Bermuda Triangle"? This seems to be what the author John Keel was suggesting. Ron Sefton, the scientific leader at Byrd Station, made light of aliens abducting Disch, but one might say that it is curious that he would bring the subject up.

Perhaps it was a planned meeting between Disch and a craft that was sent from another nearby base to get him. Disch was declared dead and has an empty cemetery plot in Monroe, Wisconsin. Is it possible that Disch is indeed still alive? If so, what strange story would he have to tell us? In 1965 the Nazi Third Power was still

A 1960 photo of sleds and a helicopter at Byrd Station.

active in many parts of the world, including Antarctica, it would seem. Had Carl Disch somehow joined the Nazi Third Power? If so, perhaps it was with the blessing of the US military. We will probably never know.

Strange Things in Argentina

If Carl Disch was taken aboard a UFO in Antarctica in 1966, he might have been taken to Argentina and one of the secret German bases or ranches located there. Argentina, as discussed earlier, was right from the start a place where submarines and UFOs were regularly seen.

The account related earlier of Captain Julian Ardanza of the Argentine ship *Naviero* seeing a "huge glowing cigar" in the water off the coast of Brazil is just one of many UFO reports from Argentina in the 1960s and 1970s, showing us that German activity was continuing in Antarctica and South America.

Let us look at UFO sightings in Argentina, as they continue to this day. The first UFO report in Argentina apparently came in 1949. This was in a town called El Maitén in Chubut province, in northern Patagonia—the region of Argentina closest to Antarctica. In 1865, Welsh people came to Chubut and settled in the Chubut Valley area, which became one of the most prosperous provinces in Argentina.

El Maitén started as a rural community but was influenced greatly by the arrival of the General Roca railroad to the area in 1939, on a branch that continued to Esquel. This branch was completed and opened in 1945, and El Maitén was selected as the site of its maintenance sheds and locomotive warehouse.

On February 20, 1949, witnesses in the town said that a flying saucer descended from the sky and landed in the town itself. After the flying saucer landed in the town, three men walked out of the craft. After a short time they returned to the craft and it took off, ascending up into the sky and out of sight.

The town at the time barely had a police brigade, a guard station, a railroad shed beside the narrow platform and a half-finished house. The town also had some small farms in the vicinity. It would seem that this small town along the Andes near the border with Chile was deemed a safe place to land a craft and have a look around. Perhaps someone was waiting for the craft in this remote town, or the occupants had simply stopped their craft here for other reasons. Was this a Haunebu or Vril craft occupied by Germans that had not surrendered at the end of war? This seems likely.

Argentina's next big UFO encounter that was promoted in the media was the sighting of a flying saucer in April of 1950 in Resistencia, a city in the northeast, very near Paraguay. The newspaper *El Nacional* of Resistencia reported on April 18, 1950:

> Around noon today, the people of Resistencia, without exception, looked to the sky to see a "flying saucer" that according to some, vanished toward the southwest in the horizon, after tracing a curve in the sky. According to others, however, it was nothing more than Venus, which was a great cause for conversation in Paraguay a few days ago. The fact is that the widest variety of conjectures and theories was put forth about this subject. The comments were the talk of the day in every meeting.
>
> Not a few people awaited the landing of the famous disk after what was reported by *El Nacional* regarding the event that occurred in Patagonia. But the "disk" went away, leaving doubts in its wake and a knowing smile on the mouths of incipient local astronomers.

One has to surmise that this is probably a German disk, very likely a Haunebu.

One of Argentina's best-known UFO cases occurred on July 3, 1960, and the witness was a high-ranking officer of the Argentinean Air Force (AAF), who by a fluke was able to obtain a remarkable photograph as supporting evidence of his encounter. The case was not given much publicity until 1977, when it finally was reported in South America UFO magazines, which are popular on that continent.

The witness in this sighting is Hugo F. Niotti, then a captain of the AAF assigned to the Air Force School for Sub-officers located in the city of Cordoba. Contrary to what many would expect, his involvement in the case did not affect his military career, and seventeen years later, when finally interviewed by a UFO magazine in Argentina, he had risen to the high rank of vice-commodore, occupying a responsible position within the AAF.

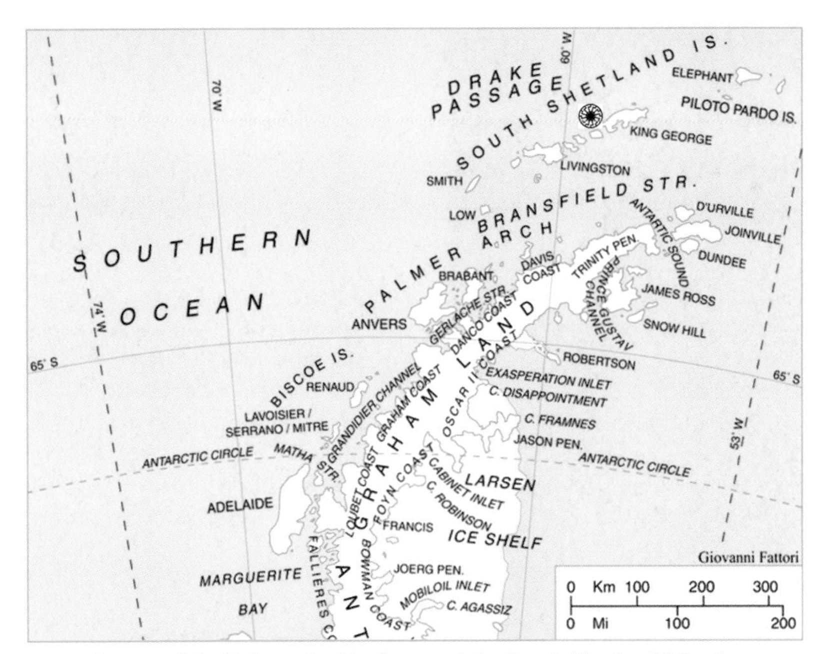

A map of the Palmer Archipelago and the South Shetland Islands.
The Black Sun logo denotes the suspected submarine base at King George Island.

An alleged photo of the mysterious "Black Knight" satellite.

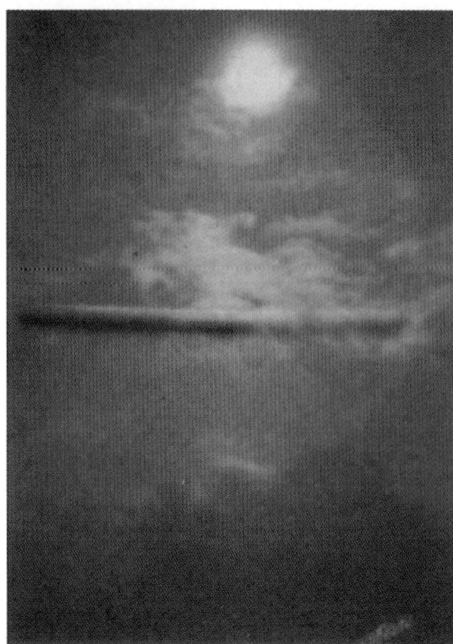

Left and below: Rare color photos of the Andromeda craft in flight. These photos were part of the 1989 document dump to Ralf Ettl. These craft were submarines and aircraft. How many were built after the war we simply don't know.

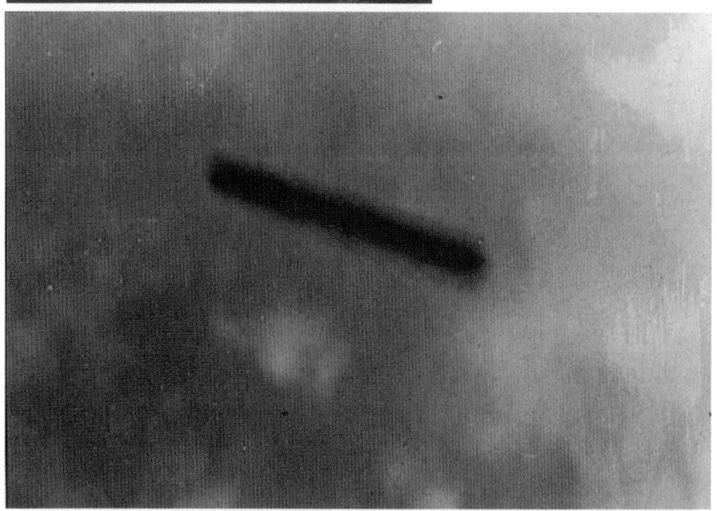

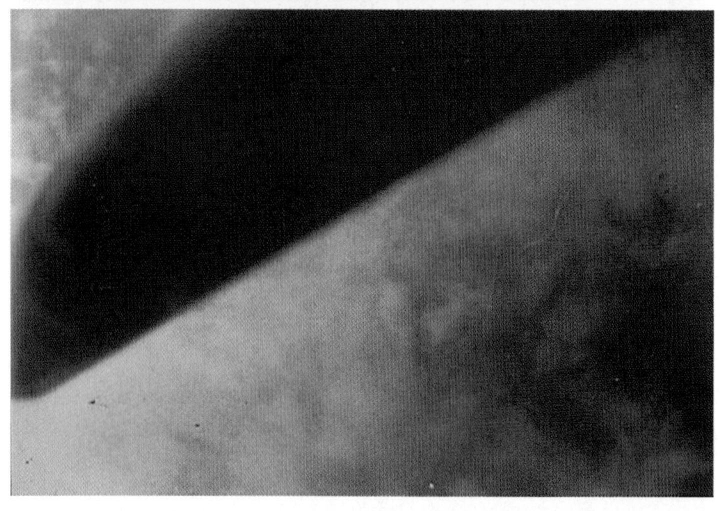

Above: A screen shot of a Haunebu at Cerro Uritorco from a television newscast. *Right*: The 1976 Argentine book *Historia de los Platos Voladores en la Argentina* (*The History of Flying Saucers in Argentina*). This book has never been published in English.

A photo of a flying saucer over Cerro Uritorco in northern Argentina.

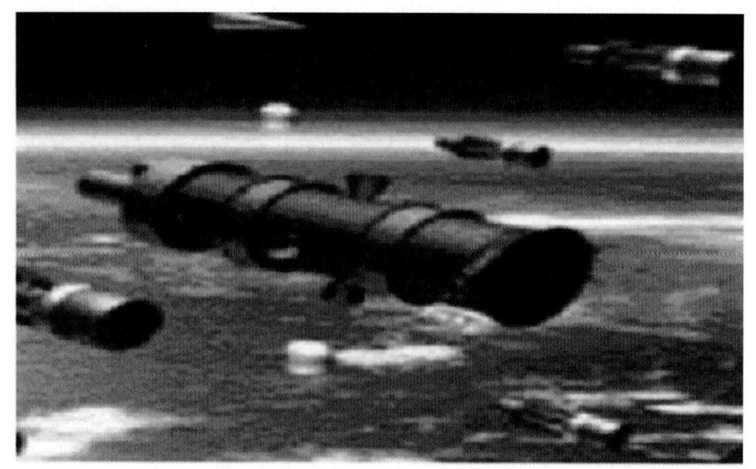

An alleged photograph of a group of US Navy spaceships from Solar Warden.

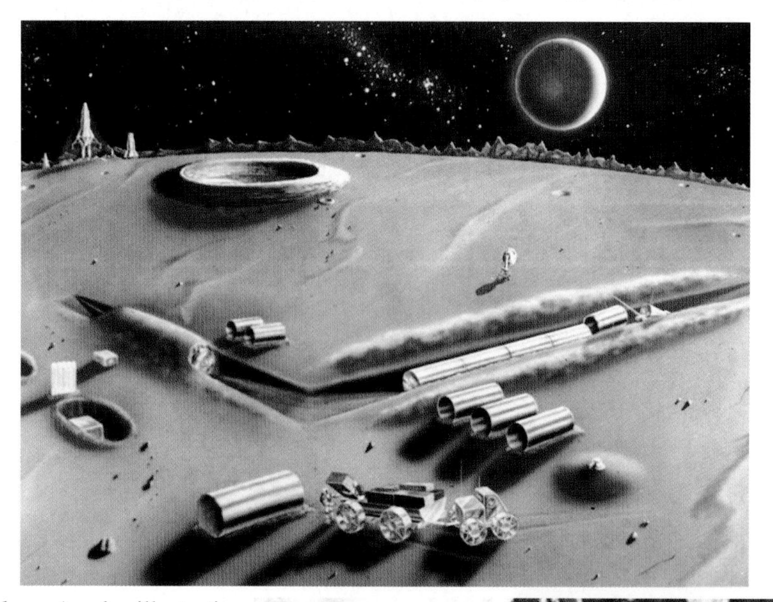

Above: A color illustration done by the US Army to show what a Moon base would look like during Project Horizon. *Right*: A curious photo from the Internet apparently showing an airvent or entrance to a secret base in Antarctica.

A map of West Antarctica showing the locations of McMurdo Station and the Beardmore Glacier.

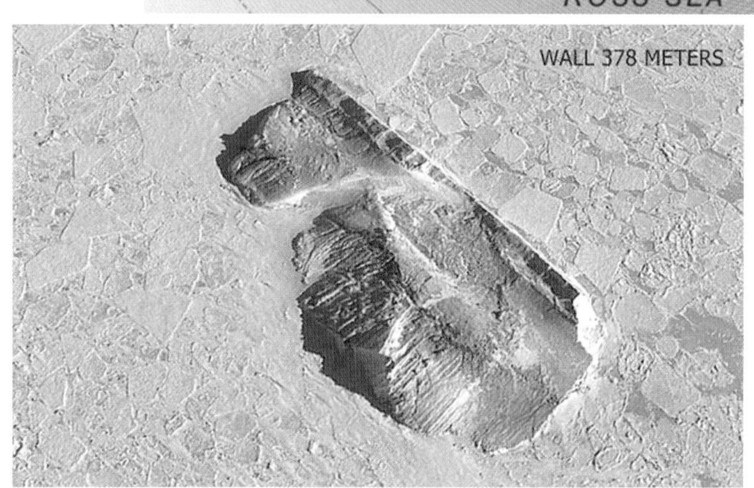

A strange wall, 378 meters long, somewhere in Antarctica, spotted on Google Earth.

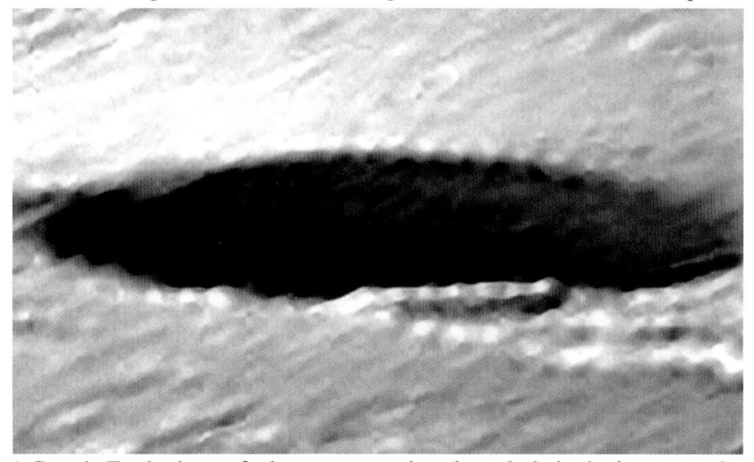

A Google Earth photo of what appears to be a large hole in the ice somewhere near the South Pole—or is it a large circular craft on the ice?

Above: An aerial photo of the Amundsen-Scott South Pole base. *Right*: A tunnel going through the ice to the dome at the Amundsen-Scott South Pole base. *Below*: Apollo 11 astronaut Buzz Aldrin preparing to fly to Antarctica and then the South Pole and on a stretcher going to a hospital in New Zealand. Aldrin is the oldest person to have visited the South Pole according to records.

A curious photograph from the Internet of what appears to be a pyramid in Antarctica.

The Google Earth photograph of a rectangular iceberg in Antarctica mentioned in April of 2017.

The mysterious lights seen in the South Atlantic from space in April of 2012.

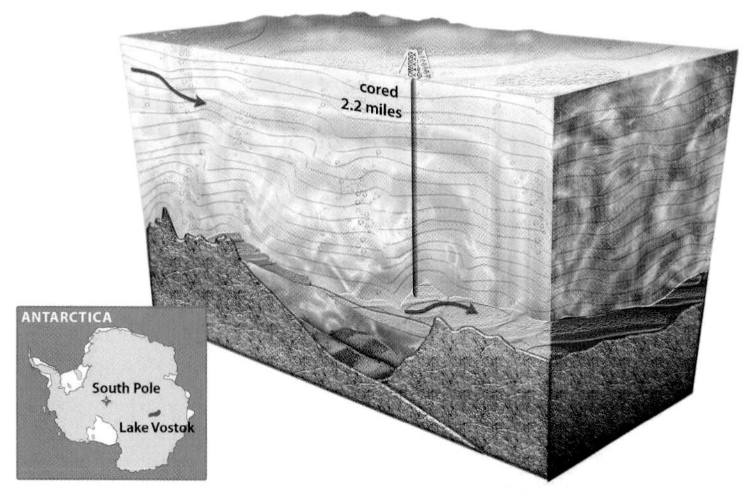

An illustration of the drilling project at Lake Vostok.

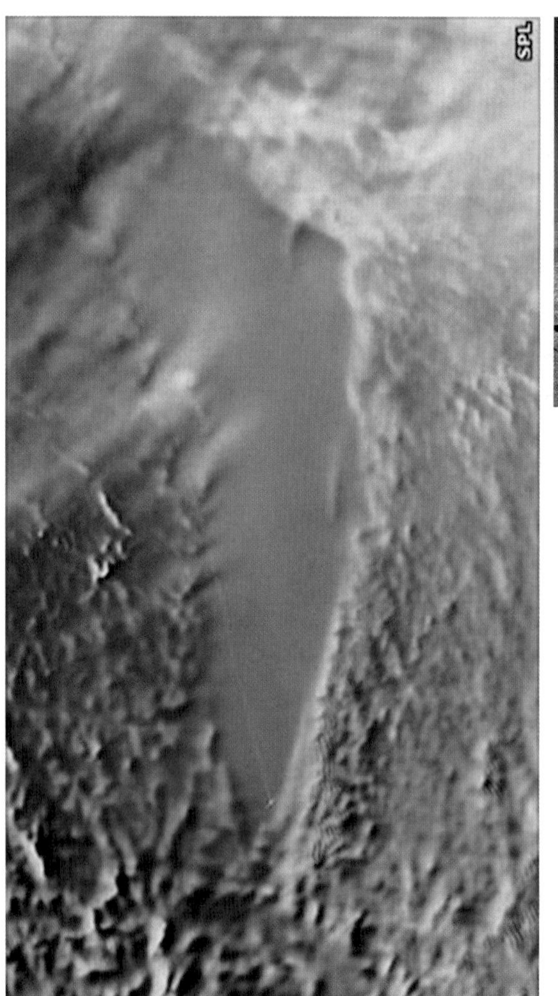

A satellite photo of Lake Vostok.

The South Pole is marked by this sign.

On July 3, 1960, then Captain Niotti was driving in central Argentina from the town of Yacanto toward Cordoba, Argentina's second largest city. After Buenos Aires, Cordoba is the most important city in Argentina, with a large population and an industrial base. In 1954 the city became the site for Argentina's automakers, which included German companies like Volkswagen and Mercedes. It is likely that Cordoba, with automakers and other industries creating and importing all sorts of mechanical parts, was a source of the material needed by the secret German factories. Some of these factories could have been located in Patagonia or Chile, but some could have been in the busy city of Cordoba itself. Indeed, this north-central city was well connected to other South American cities and would be a good place to have a covert business or two.

On the day of Captain Niotti's encounter, there was a slow drizzle and the road was rather slippery. At about 4:30 in the afternoon Niotti had just finished negotiating a wide S-curve, when

Hugo Niotti's July 3, 1960 photo taken near Cordoba, Argentina.

he suddenly noticed a rather close and unusual object hovering near the ground to the right of the road.

Startled, he stopped the car, grabbed his camera (fortunately next to him on the seat), got out and moved a few steps away, and proceeded to take a photo of the object, which was moving slowly. While he was engaged in winding the film to take a second shot, the object started to accelerate and disappeared into the clouds, which were very low. Captain Niotti jumped back into his car and continued his trip to Cordoba, where he proceeded to have the film processed.

Niotti said the object was conical in shape, with a height of seven to eight meters and a base diameter of three to four meters, with its axis almost parallel to the ground and its base facing the witness. It was at a distance of 80 to 100 meters from his location and moving very slowly toward the south, always parallel to the ground. The craft rotated very slowly. It then accelerated very rapidly, and disappeared into the low cloudbank. This sudden acceleration without any sound was inexplicable to the witness in view of his proximity. The color of the object was a uniform dark gray. He said the surface was perfectly smooth without joints or rivets and had a definite metallic aspect. A horse can be seen in the photo, turning to look at the object.

The city of Cordoba is near the famous UFO town of Capilla del Monte, mentioned in an earlier chapter and coming up again shortly.

The Volga Germans

More UFO encounters occurred throughout the 1960s, including an incident on May 12, 1962 when three truck drivers, Valentino Tomassini, Gauro Tomassini and Humberto Zenobi, were travelling on Argentina Route 35 from Bahia Blanca to the town of Jacinto Aráuz.

At 4:10 in the morning they saw an object on the ground in a field next to the road at a distance of about 100 meters from them. The object looked like a railroad car and was illuminated. As their truck came close to it, the object rose up and crossed the road at a height of about four meters. The craft's lights then went out and from the lower part of the craft came a reddish flame. As the craft began to take off the object suddenly divided into two parts, and the two parts flew off in different directions. The men were stunned and went immediately to the police department and local newspapers.

That this strange encounter happened just near the town of Jacinto Aráuz is quite curious, because this is no ordinary town in Argentina, this is a town that was founded and populated by Germans in the 1930s. They are known as Volga Germans.

The Volga Germans are ethnic Germans who colonized and historically lived along the Volga River in the region of southeastern European Russia around Saratov and to the south. Recruited as immigrants to Russia in the 18th century, they were allowed to maintain their German culture, language, traditions and churches. They were a range of Christian denominations: Lutheran, Reformed, Catholic, Moravian and Mennonite.

They even had their own independent republic within the USSR called Volga German ASSR. It had a southeastern border with Kazakhstan. During World War II, after the German invasion of the Soviet Union in 1941, the Soviet government considered the Volga Germans potential collaborators, and deported many of them eastward, where thousands died. The Volga German Autonomous Soviet Socialist Republic was dissolved. It existed from 1924, shortly after the USSR was formed, to 1941.

The Deportation of the Volga Germans was the Soviet forced transfer of the whole of the Volga German population to Siberia and Kazakhstan on September 3, 1941. With secret orders, the trains were to stop at a station just outside of the region and the adult men were to be separated from their families. They were sent to Siberia or other areas to work in labor camps during the war. Of all the ethnic German communities in the Soviet Union, the Volga Germans represented the single largest group expelled from their historical homeland. After the war, the Soviet Union expelled a moderate number of ethnic Germans to the West. In the late 1980s and 1990s, many of the remaining ethnic Germans moved from the Soviet Union to Germany.

However, a number of Volga Germans had emigrated to Argentina before WWII. In the early 1930s, following several years of drought in Europe, Volga German families emigrated to Argentina, Chile and Brazil. Many Volga German families who had settled earlier in the rural areas of La Pampa Province resettled to towns and villages like Jacinto Aráuz in search of work. This is the area of Argentina just south of Buenos Aires, with the coastal port of Bahia Blanca as the most important town. The town of Jacinto

The Volga German Autonomous Soviet Socialist Republic, 1924-1941.

Aráuz is inland, to the northwest of Bahia Blanca.

So, the town of Jacinto Aráuz is not just an ordinary town in Argentina, it is a town known for its large German population where many people speak German, rather than Spanish. This would be a good town for Nazi war criminals to inhabit and a good town to recruit new German-speaking personnel to continue the secret struggle of former officers of the Third Reich. We might imagine that certain young men from Jacinto Aráuz may have ended up at the Antarctic bases in Neuschwabenland and the South Shetland Islands.

It is also interesting to note that Patagonia does have an indigenous native population that may have also been recruited on occasion by the breakaway civilization that was becoming the Third Power. If so, these Native American young men with dark hair and dark complexions may have been part of the crew on some of the Haunebu and other craft. Perhaps this is the reason that some UFO reports, and Men in Black reports, talk about brown-skinned oriental-looking men being the occupants of such craft.

With the Volga Germans in South America and their harsh treatment in the Soviet Union, it is easy to see how the surviving elements of the Third Reich deemed Russia to be their major enemy and saw the United States as an ally—as long as it was opposed to the Soviet Union. The postwar German influence in many South American countries is well known and much of its focus was on supporting fascist military dictatorships and anti-communist movements. We also see here the beginning of a détente between the surviving Germans and the USA. With Rheinhard Gellen's Project Paperclip scientists now in top positions at NASA and Redstone Arsenal in Alabama, the German integration into the secret space program could cautiously move forward.

More UFO Sightings

More curious UFO sightings continued in Argentina in the 1960s. On December 1, 1965 starting at 8:30 pm, calls began arriving at the privately-owned Adhara Observatory in the town of San Miguel, in Buenos Aires province. Astronomers at the observatory received several calls concerning disk-shaped objects visible in front of the Moon. The staff of the observatory began photographing the Moon in fixed intervals and after processing, some of the photos revealed disk shaped objects flying in front of the orb.

This curious but brief account makes us wonder if these disk shaped objects, seen from Argentina, were actually coming from the Moon or just flying in front of a bright Moon, possibly for viewers on the ground to witness. It is intriguing that the craft might actually be going to or coming from the Moon—perhaps a Nazi base—or an American one?

On December 4, 1967 Chilean and Argentine naval ships were watching a volcanic eruption on Deception Island in the South Shetland Islands off Antarctica when one of the officers took a photo of a flying saucer that was also apparently watching the volcanic eruption. This photo is included in Wendelle Stevens' 1987 book *UFO Photos Around the World, Vol. One*,[59] and we reproduce it here. This was the only photo taken of the flying disk as it hovered near the ash plume that everyone on the ships was looking at.

Deception Island is in the South Shetland Islands archipelago, known for being one of the safest harbors in Antarctica. This island was also the base for the British Operation Tabarin discussed in earlier chapters. This island is the caldera of an active volcano, which erupted and seriously damaged scientific stations there in 1967 and 1969. Deception Island is only a short distance from King George Island that has Admiralty Bay, which was the site of the 1961 incident of a UFO coming out of the ocean, scattering frozen ice, and flying away into the sky. A lot seems to happen around these islands.

The incident at Deception Island in 1967 is described briefly in Wendelle Stevens' book by Argentine UFO researcher Sigurd von Wurmb, who speaks in the third person:

> On 4 December 1967, Chilean and Argentine Naval ships were standing off Deception Island in the Antarctic, evacuating personnel with helicopters as rapidly as they could be found. A massive volcano was erupting from the depths of an old lake on the island and the large-scale rescue effort had immediately begun. Some 30 Chilean scientists, 14 Argentines and 8 Brittanos were recovered as tons of ash, rocks, and moulten lava were thrown high into the air.

The photo of a flying disc taken December 4, 1967 at Deception Island.

Enormous masses of lava and incandescent material was spewed from the fissures, changing the landscape of the island. A 300-foot high volcanic cone was eventually raised.

A photograph of one of the massive eruptions, taken by a Chilean ship's photographer showed a high domed disc-shaped object in the air in the upper left hand corner, looking toward the erupting mass. The photographer, occupied with the rescue effort when he snapped the photo, did not remember seeing the object when he took the shot and was surprised to see it on the developed print. Sr. Sigurd von Wurmb obtained this copy from a Navy Admiral.

A copy of this photo was published in *La Chronica* of Sante Fe, Argentina, dated 24 October 1970. It was furnished by Comodoro Palma of the Argentine Navy.

This photograph was taken near the end of two years of heavy UFO activity over the Antarctic. All of the Antarctic stations reported them, and several other photographs were also taken. Two huge cigar-shaped craft were observed and

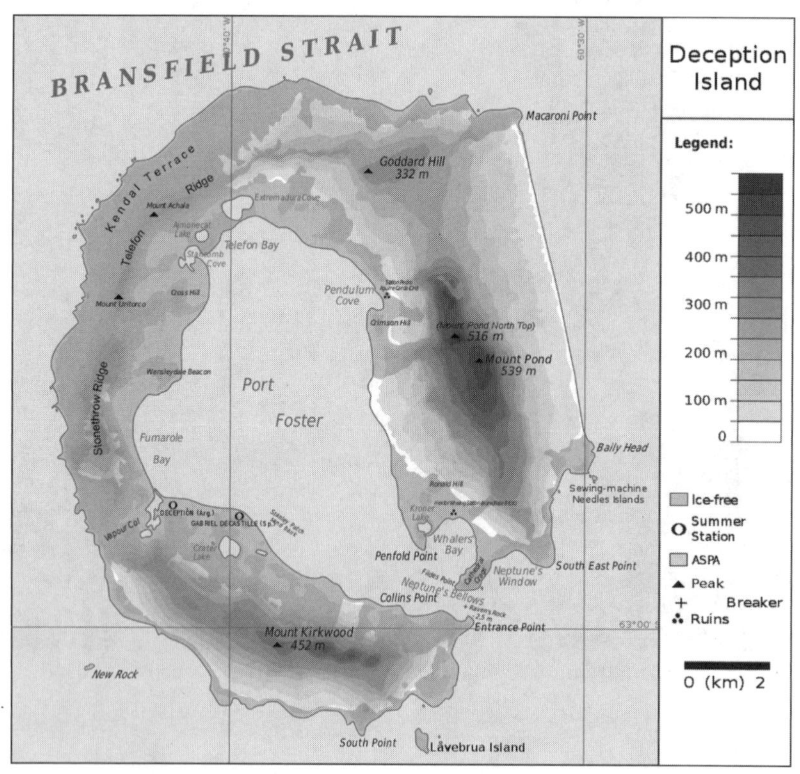

A map of Deception Island in the South Shetland Islands.

photographed maneuvering over and about the area for two days during this time. —Sigurd von Wurmb

Well, lots of fascinating material to pick over for a moment. We learn from Sr. von Wurmb, apparently a German-Argentinian, that the photograph was released by an Admiral in the Argentine Navy, though it took a few years to get published, and only then in a small newspaper in southern Argentina. It did not make it into an English publication until 1986, when Wendelle Stevens published his UFO photos books in two volumes. This is one of the few UFO photos ever taken in Antarctica, though von Wurmb tells us that there are other photographs of UFOs taken during this time.

In fact, he gives us the very intriguing information that two cigar-shaped craft—submarines that can fly—were seen over a two-day period during 1967. This seems to confirm that the surviving German base in Antarctica was in the South Shetland Islands, just off the coast of Antarctica, possibly on King George Island where Admiralty Bay is also located. All of the South Shetland Islands have fjords and massive granite cliffs coming out of the water. There are many good spots in these islands for a secret submarine base. Deception Island has an active volcano, and other geothermal activity supplies hot springs and even geothermal power. A secret U-boat base in the South Shetland Islands could even have hot springs and baths inside it. Just to be clear, the South Shetland Islands are not in the area of Neuschwabenland.

The 1970s in Argentina saw the famous Dionisio Llanca case and also saw the rise of a popular media and UFO personality named Fabio Zerpa.

The "Dionisio Llanca Case" occurred on October 28, 1973 in the vicinity of Bahia Blanca in the province of Buenos Aires, Argentina. The case gained worldwide fame. The witness, truck driver Dionisio Llanca, said that he was changing a flat tire around 1:30 in the morning when two men and a woman, dressed in tight-fitting grey outfits and yellow boots and gloves, appeared out of nowhere beside him. He felt paralyzed and he could see their craft suspended some seven meters in the air, over some nearby trees.

He described the craft as measuring some four meters in diameter, and being completely silent. Before this, he had only noticed a yellow light approaching along the highway. He was able to hear the beings talking among themselves in an unfamiliar language until one of them lifted him by the neck while the other pinched his one of his fingers. The last thing he saw was two drops of blood sliding from his index finger to the ground, and the woman's fixed gaze. He then became unconscious.

He woke up sometime after three in the morning to find himself lying on the ground beside some railway boxcars on the premises of the Sociedad Rural, a country building for civil meetings and gatherings, similar to a grange, 10 kilometers away from where his truck was parked. The vehicle was collected by police officers later that afternoon.

Dionisio says he got up and began walking down the road, unable to remember his own name or how he had gotten to this place. He then obtained help from a man—whose identity remains unknown—who led him to the local police station. When he tried to speak with the police he was having such difficulty that they thought he was drunk and simply released him back into the care of the stranger. This stranger then drove him to a hospital where he spent the next several days.

The case became widely known thanks to journalists from two Buenos Aires newspapers who were correspondents in the city at the time. But the story did not end there. As the strange event circulated through the country's print media, renowned researcher Fabio Zerpa arrived on the scene, along with reporters from such magazines as *Gente* and correspondents from ufological publications seeking to interview the abductee.

The case was investigated by Fabio Zerpa and his findings were published in his 1990 book *El Reino Subterráneo* (*The Underground Kingdom*). Fabio Zerpa was born in Uruguay in 1928 and died in Argentina on August 7, 2019. He moved to Argentina in 1951 and was an actor, parapsychologist, radio host, author and UFO researcher.

Zerpa had a short theatrical career and became increasingly interested in UFOs and extraterrestrial life, having already studied psychology. After some years of investigation, Zerpa started to host his first conferences in the early 1960s, and then in 1966 he created a popular Argentine radio program called *Más allá de la cuarta dimensión* (*Beyond the Fourth Dimension*). Zerpa reported on more than 3,000 cases of UFO sightings and in every one of them he spun the angle that the craft were extraterrestrial.

Among his many books are such titles as: *El OVNI y sus Misterios* (*The UFO and Its Mysteries*), Spain, 1976; *Dos Científicos Viajan en OVNI* (*Two Scientists Travel by UFO*), Argentina, 1977; *Los Hombres De Negro y los OVNI* (*Men In Black and UFOs*), Spain, 1977; *El Reino Subterráneo* (*The Underground Kingdom*), Argentina, 1990; *Los OVNIs Existen y son Extraterrestres* (*UFOs

Fabio Zerpa in 1969.

Exist and They Are Extraterrestrials), Argentina, 1995; and more. Zerpa's books were full of interesting stories and one thing that we might get from his books is that there are a lot of UFO sightings and tales of the Men in Black in Argentina and South America. Go figure!

His curious book on the Dionisio Llanca affair proposed that the aliens that took the truck driver aboard their craft were coming from an underground world, similar to the hollow earth, from whence these flying saucers and articles of clothing came. Maybe this underground world was in Antarctica. In Zerpa's world of the fantastic, anything was possible. For decades he was the face of ufology in Argentina and other countries like Chile and Uruguay. In some ways Zerpa steered the public away from seeing any German or WWII dimension to all of the UFO sightings in Patagonia and elsewhere—he turned it all into a whirling fantasia of interdimensional weirdness of extraterrestrials and the inner earth.

Zerpa conducted hypnotic regression sessions with Llanca, and during them the truck driver said he ascended into the craft on a beam of light. He said the craft had only one window, many levers, several monitors and a radio, which aside from speaking in Spanish, repeated that they were friendly to humans and had come from a secret location. Before releasing him, they promised that they would find him again someday.

Llanca also described a detail that gave some credibility to his story. He said that while he was aboard the flying saucer, two hoses were deployed from the object: one of them made contact with a puddle of water and the other with a high-voltage line. Curiously, at that time the city experienced a major power loss that was felt in various neighborhoods. Had the power outage been caused by the flying saucer? Llanca says yes.

In a 2013 interview in an Argentine newspaper Llanca is quoted in a lengthy story:

> "In 1976, tired of being injected with sodium pentothal—the so-called truth serum— and the problems that remembering the event were causing me, I decided not to appear again," he says. His rush to obscurity took him places he never imagined. During the next stage, he spent

> two and a half years interned in hospitals in Mendoza, Buenos Aires and Rawson. "I was shedding skin all over my body, and my eyes became so red at times I thought I might be going blind. The doctors told me I showed signs of radiation exposure, although worst of all is that I would see the light from the UFO every so often," he says.
>
> Even today, whenever he washes his hands, he notes that a small puncture they made when they injected him for a second time is still visible. This happened on the craft, while a sort of loudspeaker spoke to him in Spanish, assuring him that no harm would befall him.
>
> He also claims to have a mark on his left eyelid, which becomes more visible when he washes his face and sees himself in the mirror. This was done with a glove that had something like thumb tacks on its surface. "These are details I remembered through hypnotic regression sessions to which I was subjected shortly after the event," he adds.

One has to wonder if these two men and one woman were aliens from inside the earth as Zerpa concluded. In light of what we know now, from the 1989 document dump and further research, it would seem that some sort of odd plan was suddenly hatched by this trio while flying their Vril craft in the vicinity. We might note their uniforms again briefly: they were dressed in tight-fitting grey outfits with yellow boots and yellow gloves. Might this be what is worn at the secret base in Antarctica?

Also, what of the strange language that they spoke? Llanca the truck driver probably did not speak any language other than Spanish. Still, he might have recognized if the three occupants of the flying saucer were speaking a language such as English, French, Portuguese or German. What language were they speaking? It might have been a language that Volga Germans would have known which would include Russian, Kazakh, Ukrainian and several older German dialects. Tibetan is another possibility. Remember, these were not three-foot tall grey aliens, they were normal-looking humans with yellow boots and gloves.

Llanca's being brought up into the craft by a beam of light in 1973 is similar to the famous Travis Walton case near Heber, Arizona in 1975. Also, was he really given sodium pentothal during his hypnotic sessions? This seems like a bit of an overboard technique, one that smacks of Nazis and the Men in Black. Was Zerpa some strange inner-circle player who played his own mind control games with Llanca and the public? In his 2013 interview Llanca virtually talks about escaping from Zerpa and Buenos Aires and going into hiding for decades in small towns in southern Argentina.

As we continue to see, starting with WWII things got very, very strange in Argentina and other parts of South America.

Close Encounters Around the World

During the 1980s in Argentina the UFO activity centered around the previously mentioned northern city of Capilla del Monte. As mentioned in an earlier chapter, Capilla del Monte is a small city in the northeastern part of the province of Córdoba, where the main tourist attraction in the area is the Cerro Uritorco, a small mountain only three kilometers from the city, famed around Argentina as a center of paranormal phenomena and UFO sightings. One famous incident happened on January 9, 1986. On that day a flying saucer was seen near Capilla del Monte by hundreds of residents. The astonished crowd watched the craft land on a nearby hill known as El Pajarillo. Later, residents rushed to the scene where they found a mysterious footprint on the ground where the flying saucer had landed. As we reported in an earlier chapter, the area around Capilla del Monte has been a major UFO area since the 1950s. There has also been a lot of activity in the area by "Men in Black." It seems likely that both of these mysterious occurrences are related to postwar German activity in the area.

About three years later on the sunny afternoon of December 26, 1988, a silver UFO flew over the northern Buenos Aires suburb (barrio) of Villa Urquiza. It is considered the most spectacular UFO incident to occur in Argentina, with more than 7,500 witnesses. The local airport reported an object flying to the west, towards General Paz Avenue, seen on the radar.

Let us pause for a moment and look at another famous UFO case of the Antipodes. This is the 1966 Westall High School UFO incident that happened in broad daylight in a suburb of Melbourne, Australia. On April 6 of that year, around 11:00 am, for about 20 minutes, more than 200 students and teachers at two Victoria state schools allegedly witnessed an unexplained flying object—looking like a flying saucer—which descended into a nearby open wild grass field. The grass field was adjacent to a grove of pine trees in an area known as The Grange (now a nature reserve). According to reports, the object then ascended in a northwesterly direction over the suburb of Clayton South.

Here is a brief description of the incident from Wikipedia:

> … a class of students and a teacher from Westall High School (now Westall Secondary College) were just completing a sport activity on the main oval when an object, described as being a grey saucer-shaped craft with a slight purple hue and being about twice the size of a family car, was alleged to have been seen. Witness descriptions were mixed: Andrew Greenwood, a science teacher, told *The Dandenong Journal* at the time that he saw a silvery-green disc. According to witnesses the object was descending and then crossed and overflew the high school's south-west corner, going in a south-easterly direction, before disappearing from sight as it descended behind a stand of trees and into a paddock at The Grange in front of the Westall State School (primary students). After a short period (approximately 20 minutes) the object—with witnesses now numbering over 200—then climbed at speed and departed towards the northwest. As the object gained altitude some accounts describe it as having been pursued from the scene by five unidentified aircraft which circled the object. Some described one disk, some claimed to have seen three.

This is an astounding incident, one that was basically covered up by the authorities as quickly as possible. The children and teachers were asked not to speak very much about the incident, and Wikipedia says that witnesses and researchers were surprised when *The Sun News-Pictorial* (a tabloid) ran no story, yet *The Age* (a broadsheet, normal newspaper) did. Other newspapers ignored the story as well and *The Sun* and *The Herald* newspapers, while not mentioning the Westall incident, both published cartoons in the following day's

editions that made light of the flying saucer phenomenon.

The one photo of the incident is a very good photo that appears to depict a large Haunebu flying saucer. This 1966 photo does not appear to be faked—and with an incident in broad daylight with over 200 witnesses, no one needs to fake a photo like this. It is interesting to note that two Australian UFO groups studied the

SAUCER MYSTERY: SCHOOL SILENT

What was it ?

CLAYTON.— After more than a week of investigation, mystery still surrounds the reported sighting last Wednesday of a flying saucer near the Westall High School, Clayton.

Investigations of the report have been hampered by the reluctance of school authorities to permit interviews with eye-witness students and staff members.

Several children attending the school and at least one staff member are reported to have seen the unidentified flying object.

They are believed to have given corroborating descriptions of the object as "a round humped object with a flat base being circled by what appeared to be light aircraft."

It was described as grey or silver grey in color, but it is not known whether the object appeared high or low in the sky or for how long it was observed.

Investigators call for eye-witnesses

Army report

art to 28

Above: An Australian newspaper article about the Westall High School incident in 1966. *Opposite:* An alleged photo of the Westall flying saucer.

incident intensely, as did the skeptic societies of Australia. None concluded that the incident was a hoax and the only suggestion that the skeptics were able to come up with was that it was an experimental military plane (sound familiar?) or a target dirigible being towed by a small plane.

So, what are we to make of this incident? Let us put aside any theories that a reptoid craft from Mars had stopped by to pick up some children for their slave labor factories underground. Rather we seem to have an incident that was meant to garner as much media attention as possible. It was a stunt—but not a hoax—of a power, the Third Power, to show off some of their aircraft to the world. It was a "hey, look at me" moment that involved landing a Haunebu on a sunny April morning near a high school in the very modern city of Melbourne, Australia.

It was staged so that the maximum number of people would witness the incident, adults and children, so that as much chatter as possible would be created and the media would have to take some notice and report it. Maybe even photos of the event would appear in newspapers and a larger section of the population would doubt what their government was telling them about flying saucers. But, sadly, the incident was largely suppressed outside of Australia, and even there major newspapers would not run the story. While many people in Western countries believe that our media is fair and free from government intrusion, this is sadly very wrong. It has been speculated that the *New York Times* ("All the news that's fit to print") has never run a story concerning UFOs except for feature stories concerning conferences or films.

Had this Haunebu made the flight from Antarctica to Melbourne? After its mission to shake up the Australian government and media did it then return to its base in Antarctica? Perhaps after that it made a mission to South America.

This Haunebu may have come from another hidden UFO base in the Australian region. It has been suggested that there are secret UFO bases on the island of New Guinea, for instance. This island is divided into two political divisions and is largely devoid of roads or towns. The interior of New Guinea could harbor a remote mountain base that would be served only by air. This would explain some of the UFO activity that has been reported on the island.

Another possible location for the UFO base is the Solomon Islands, where a lot of UFO activity has been reported. There are other South Pacific islands such as Vanuatu or New Hanover where secret UFO bases may be located as well.

In fact, since these craft are able to fly "from pole to pole" as Byrd said, they turned up all over the world, sometimes hovering in broad daylight over major cities. In his book *The CIA UFO Papers,*[40] investigator Dan Wright says that CIA reports from 1953 indicate that there was a great deal of UFO activity around the world that year. He says a document refers to the Danish Defense Command remarking that the "flying saucer traffic" over Scandinavia was of immense aero-technical interest.

Another CIA paper dated August 18, 1953 conveyed accounts of anomalies from newspapers in Athens, Greece; Brazzaville, Congo; and Tehran, Iran. Wright then mentions another CIA paper from that year that said:

> …a German engineer claimed that flying saucer plans, drawn up by Nazi engineers before World War II's end, had come to be in Soviet hands. The source claimed German saucer blueprints were already underway in 1941. By 1944, three experimental models were ready, one in disc shape. All could take off vertically and land in a confined space. After a three-month siege of the German's Breslau (now Wroclaw, Poland) facility at the war's conclusion, Soviets stole the plans on saucer construction.

Wright then describes another curious CIA document with an account from November 22, 1952:

> A missionary and five companions in French Equatorial Africa had had a close encounter. Driving at night, they had witnessed four motionless discs overhead that lit up like suns when in motion but were silvery when stationary. Over 20 minutes the four moved about the area, seemingly performing tricks, then hovered momentarily before leaving non-uniformly. Later the six witnesses saw four objects forming a square at cloud level. One lit up vivid red and rose

vertically; the other three joined it to form a square again. Luminous aerial objects were seen in the same time period above Homs, Syria, and the oil fields at Abadan in west-central Iran.[40]

Wright mentions two curious cases in Africa from 1952 in the CIA documents. One incident took place on June 1, 1952 when the master and first mate of a cargo ship just off of Port Gentil, in the West African country of Gabon, witnessed an orange luminous object rise up behind the port, do two right-angle turns, pass overhead, and continue out of sight. Was this a Haunebu picking up or letting off passengers in a field behind the port?

This craft might have then departed for Laâyoune (or El Aaiún) that at the time was the capital of the colony of Spanish Sahara. It became a modest city in the 1930s and in the year 1940, Spain designated it as the capital of Spanish Sahara. Because Spain was a neutral country, but pro-Nazi (like Argentina), its colonial lands were used as proxy territory by the Third Reich, and U-boats and aircraft made routine stops at locations where supplies were available and transactions could take place. The country was disestablished in 1976 and today is basically part of Morocco. One early report said that Hitler was living in Laâyoune immediately after the war.[33]

Wright then discusses a CIA document on UFO activity in July of 1952 in Algeria. Starting on July 11 in Lamoriciere, Algeria, several UFOs were seen including a longish, fiery oval. Then four days later a flying saucer was seen at the town of Boukanefis. Then on July 25, at 2:35 in the afternoon, UFOs were seen over a factory and other places near the coastal city of Oran. The same night near the Algerian town of Lodi, southwest of Algiers, several UFOs were seen.

The next morning at 10:45 in broad daylight in Tiaret, Algeria, five persons saw a shining cigar-shaped UFO with a darkened center silently traversing the sky. A similar UFO was seen later that night in the Algerian town of Eckmuhl. The CIA documents say that a few days later, on July 30 in Algiers, a black disk was seen by a woman to descend from the sky and then suddenly move horizontally out of sight. Later that night, in the early hours of July 31, UFOs were seen in the sky by multiple people around Algiers, the main city of the country, and in other areas like Oued and Tlemcen. Finally, later that morning in broad daylight (11:30 AM) a couple driving just outside of the major city of Oran saw a UFO cross the sky at great speed.

Wright's fascinating book has many interesting incidents, and let us remember, they are from CIA documents that were kept confidential for decades and have only been released because of the important Freedom of Information Act. Other countries do not have such government-sanctioned legal avenues for obtaining data.

There are far too many CIA documents reviewed in Wright's book to discuss right now, but it is certainly a great resource for UFO incidents, many of which fall within the parameters of this book. A few final mentions from Wright's book are one in Baku, and one in Bahia Blanca, Argentina.

Wright says that a CIA document from 1955 refers to an incident in Baku, the capital and largest city of Azerbaijan. Formerly a city with a large population of Armenians, Baku is famous for a big battle that took place in 1942. Known as the Battle of Baku, the German Army approached within some 530 kilometers (329 miles) northwest of Baku in November 1942 in an effort to capture the oil facilities there. They fell short of the city's capture before by being driven back during the Soviet Operation Little Saturn in mid-December 1942. This was part of Germany's desperate attempt to control oilfields across Eastern Europe, as Germany had no oil fields of its own. The Battle of Baku was basically Germany's final failure to secure major oilfields during the war. Baku is also in the vicinity of the Volga Germans.

The CIA document begins by saying that in 1953, a disc-shaped UFO was seen at the Soviet city of Shakhty, a coal-mining town at a spur of the Donetsk Mountains in southwestern Russia. The document then says that on August 9, 1955 outside of Baku, Azerbaijan witnesses saw a "flying saucer" take off two miles or so in the distance. They described a glow around the craft as it rose up into the sky.[40]

So, here we may be looking at some Haunebu craft picking up special people, possibly Volga Germans, at the town of Baku and whisking them away to someplace else. The base in western Tibet is a possibility, and that base was probably still active in 1955. One might think that it was abandoned in years after that as the

Communists expanded their control over Tibet, a sparsely populated area with few roads and towns. But who knows? The Nazis would have supported the original government of Tibet, one of an enigmatic hierarchy of Lamas of various sects, typically opposed to each other. At the time of WWII and immediately afterwards, there was effectively no Dalai Lama as the previous Dalai Lama had died in 1933 and the current Dalai Lama was only a small child.

Wright mentions briefly a curious incident when a young man saw a flying saucer move silently over pastures near the Argentinian port city of Bahia Blanca on September 19, 1971. Says Wright:

> The object stirred up whirlwinds of dust, while cows appeared to change color in its presence. Afterward, his face was severely burned and he suffered from a persistent migraine headache.[40]

UFO sightings continued throughout the 1970s and 1980s in Argentina and in 1991 a major flap of UFO sightings occurred around the town of Victoria, a city located in the southwestern part of the province of Entre Ríos, just west of Buenos Aires. Victoria is located on the eastern shore of the Paraná River, opposite the major city of Rosario, which is in Santa Fe province. The Paraná River here is quite wide and commercial fishing is a major activity in the area.

Witnesses Andrea Pérez Simondini and his wife Silvia wrote in 1998:

> Upon reaching the area, we installed ourselves and our gear at the Victoria campgrounds, since it was one of the best observation spots facing the lagoon, dominating every detail in the island sector, from which the maneuvers of these objects could be witnessed constantly. At the edge of one of the canals bordering the campgrounds, we saw a mustard-yellow light toward the north, rising over the horizon and remaining stationary for some 8 minutes, as if hanging from the sky. The possibility that it was an antenna we hadn't noticed earlier was discussed, or had it really been an object? Momentarily, faced with disbelief, I stared at the

light, which after a few minutes began traveling toward the northeast, that is to say, toward the center of the city.

I began signaling it with a flashlight, and it suddenly changed its trajectory toward me. I quickly went to where the others were watching, as they had also noticed the change in movement.

When the object crossed a row of trees that runs along the edge of the camping, we noticed that it had lights facing forward. As it flew overhead, those lights cast their beams over us as though looking downward. In a matter of seconds, the object turned off and vanished as if it had never been there. Our astonishment lasted for hours—we could not believe such a thing could have happened to us. I think the story conveys the sensation that engulfed all those who were witnesses to this incredible experience.

...We interviewed direct witnesses to the objects' maneuvers, those who had found possible evidence of them, those who claimed having seen entities or beings emerging from craft that landed on the fields. A wealth of information allowed us to establish some significant statistics. But the substantial element was that these objects—contrary to what was believed—did not emerge from Laguna del Pescado. Rather, they emerged from one of the branches of the Victoria River, a brook with a strong current. This brook was the deepest area, and eyewitness accounts not only described the lights that emerged or came out of there, but also the noises that could be heard under the water. We stationed ourselves there several times and on several occasions were auditory witnesses to the sounds that caused great fear among the locals.

Regarding the objects, they said that their luminosity was so great that many of their homes had dark drapes over the windows, since the objects could bathe the ranches in light. It is somewhat humorous to see all these little houses with black curtains.

So, what of this amazing flying saucer light show over an area not too far from the capital megacity of Buenos Aires? It appears that the craft, seen coming out of the water at times, made deliberate passes over ranches and other areas with its powerful lights on full blast to illuminate people's houses in the middle of the night. Why would sneaky extraterrestrials be so overt about their activities in this area? They wanted people to know that they were there. Were they trying to frighten people? It doesn't seem like it. Were they completely unafraid of the Argentine authorities? It does seem that way. Were they trying to be noticed in order for people to realize that the flying saucer phenomenon was real, and not just a fantasy coming out of the media?

Indeed, having spent years travelling around South America, starting in the mid-1980s, I have found that people on this continent have a high degree of belief in a real UFO phenomenon. People on the street, the media, and even the governments, all largely admit that UFOs, called OVNIs in Spanish, are very real.

There is just too much UFO activity in South America to cover exhaustively here. By this time the reader should have a pretty good idea of just how strange things have gotten in Argentina, Chile, Brazil and other countries since the end of the WWII.

This UFO activity continues in the sky and in the waters around South America and Antarctica. Before we look at the South Pole base, let us look at something very strange seen in the South Atlantic in the year 2012.

An Astounding Light Show

A NASA satellite image captured an astounding light show during April of 2012 and was published in the Brazilian magazine *Revista UFO* the next year. The strange light formation that has been explained away as "maneuvers by fishing boats trying to attract fish with lights."

The photo, seen in color in the color section of this book, shows a huge spinning "wheel" of light, about 300 kilometers out into the southern Atlantic, off the eastern coast of Patagonia. It is only a few hundred kilometers from the coast of Antarctica. The huge lights look like the spokes of a massive wheel, partially underwater. Another large lit object is seen to the lower left, close to Tierra del Fuego.

The article in *Revista UFO* read in part:

> A NASA satellite detected a large and strange concentration of lights in the middle of the Atlantic Ocean, 300 km from the coast and along the southern edge of America, parallel to the Patagonian shoreline to be exact. The concentration lasted only a few days, leading to much speculation about its origin by the team of scientists monitoring the events, given the full awareness that there was nothing in the area identified that could justify the lights, such as oil rigs.
>
> The agency tendered a possible explanation: a formation of fishing vessels, grouped in mid-ocean, with lights so powerful they could be seen from space. However, it is difficult to explain that their light emissions, as shown in the photo, exceed those of the city of Comodoro Rivadavia, with a population of 137,000.
>
> The lights were seen using the Visible Infrared Image Radiometer Suite Suomi NPP satellite. According to NASA, night fishermen are after *Ilex Argentinus*, a short-finned squid. The squid is found hundreds of kilometers away from the shore near Tierra del Fuego.

One has to wonder if this is really a good explanation for these lights. Since it is an official NASA photo they would need to come up with some plausible explanation such as a huge mass of fishing boats beaming their lights in a rather unusual formation in the South Atlantic. One has to wonder if something else might be going on, perhaps even under the water. Perhaps these lights have something to do with Antarctica and the secret space program.

More recently, on April 24, 2017, the *Express* newspaper in Britain (express.co.uk) published an online story with the headline: "Shock claims Antarctica is UFO BASE for ALIENS." This headline was followed by a teaser sentence that read: "UFO hunters have long claimed that aliens have secret bases hidden in the Antarctic where they hide away from view. By Jon Austin." The curious photo of a rectangular iceberg thought to be an alien base is shown and the brief story follows:

The Google Earth photo of the odd iceberg in Antarctica.

Conspiracy theorists claim the Nazis experimented with UFO technology secretly in Antarctica. Now conspiracy theorists claim to have found evidence of an alien base on Smyley Island, off the west coast of the Antarctic Peninsula, on Google maps. They believe this peculiar object, around 500-meters long, near the Antarctic could be an underwater alien base. Theories about secret bases in the Antarctic have circulated for years—with one conspiracy detailing that the Nazis had secret bases in the southernmost continent which they used to experiment on captured UFOs during WWII. The new object was spotted on Google maps by website UFO Sightings Hotspot which has revealed its "findings."

This huge floating rectangle of ice is interesting, as icebergs are not typically rectangular. A giant iceberg could be used as a submarine base of sorts, but one can see a lot of problems with such a scenario. It seems unlikely that this an alien base, either. That would more likely be deep under water. What is most interesting is that Nazi bases and Antarctica are discussed in the story at all, showing that this whole "hidden history" is still part of our pop culture today.

Strange Activities in Antarctica

It has been alleged recently that an alien city has been found beneath the ice at the South Pole. It has also been alleged that there are pyramids peeking up out of the ice in Antarctica. It is alleged that there are entrances that can be seen going inside of icy mountains and inside of ice sheets. It is alleged that there is alien activity in Antarctica. It is alleged that Apollo astronaut Buzz Aldrin went to the South Pole to view frozen aliens. It is alleged that Antarctica houses much of the activity of the secret space program.

Some of these allegations may be true. Some may not. Let us look at some of the recent strange activity in Antarctica. On the online site for the *Express* newspaper in Britain (express.co.uk) they published this story on March 16, 2017 concerning a whistleblower who had contacted Linda Moulton Howe:

> There is a huge secret hidden beneath the ice in Antarctica. Military whistleblowers report large alien structures under two miles of Antarctic ice. Navy Seal Spartan 1 walked an alien hallway inscribed with mysterious hieroglyphs and tells investigative reporter Linda Moulton Howe that Antarctica's alien presence spans centuries, linked to Star Gate portals around Earth and beyond our solar system.
>
> UFO whistleblower: Ex-US Naval officer 'saw entrance to secret alien base in Antarctica.'
>
> A FORMER US naval officer has shockingly claimed to have seen the entrance to a secret alien base and UFOs while on duty in the Antarctic.
>
> The former military man, who is said to have seen 20 years of service, has allegedly revealed the startling story to US conspiracy theorist investigative journalist Linda Moulton Howe. Mrs Howe, 75, from Albuquerque, New Mexico, who has been awarded for her paranormal research through website Earthfiles.com, says she was contacted several times by the whistleblower she has only been referred to as Brian.
>
> In a series of videos released on YouTube, Mrs Howe is heard interviewing Brian about what he allegedly saw.
>
> Brian, 59, who alleges to be a former retired US Navy

petty officer first class flight engineer in a squadron called Antarctic Development Squadron Six, claimed to have been part of a crew that flew through a "no fly zone" above Antarctica, and saw UFOs, aliens, and a giant entrance hole to an alien base.

He was said to be stationed there between 1983 to 1997, when he retired, and on several occasions saw "aerial silver discs" flying over the Transantarctic Mountains.

In the videos, he claimed there is a top-secret collaboration between humans and aliens, with Antarctica a major research ground for the projects.

The large hole was said to be five to ten miles from the South Pole, in the supposed no fly zone.

He also said a group of scientists had gone missing for a week and come back terrified and refusing to speak.

Then at a camp near Marie Byrd Land, some dozen scientists disappeared for two weeks and when they re-appeared, Brian's flight crew got the assignment to pick them up. Brian says they would not talk and "their faces looked scared."

In an email to Mrs Howe, he said: "Another unique issue with South Pole station is that our aircraft was not allowed to fly over a certain area designated five miles from the station.

"The reason stated because of a air sampling camp in that area. This did not make any sense to any of us on the crew because on two different occasions we had to fly over this area.

"It was on the opposite side of the continent and we had to refuel at South Pole and a direct course to this Davis Camp was right over the air sampling station.

"The only thing we saw going over this camp was a very large hole going into the ice. You could fly one of our LC130 into this thing. Talk among the flight crews was that there is a UFO base at South Pole and some of the crew heard talk from some of the scientists working with and interacting with the scientists at that air sampling camp/large ice hole."

Earthfiles has published a redacted DD-214 document and Antarctic Service Medal given to Brian on November

20, 1984 in a bid to confirm his credentials.

Such reports are fascinating indeed. Linda Moulton Howe is the sort of person with a website, such as Earthfiles.com, that attracts information from "insiders" and "whistleblowers" who want to anonymously tell their strange stories. Certainly, we should listen to these stories, even if we don't know their sources. There are essentially three different stories here: the first is about silver flying disks that were seen flying over Antarctica according to the source. Then there are two groups of scientists who went missing for a time but then resurfaced. They were quiet and did not speak about their ordeal. Then there is a big hole in the ice about five miles from the South Pole. This is particularly intriguing and the inference is that this hole is an entrance to an under-ice base near the South Pole. It should be pointed out here that the ice at the South Pole is almost two miles thick.

We will return to what is at the South Pole in a moment, but let us look at an even more recent post by Linda Moulton Howe as summed up by UFO investigator and author Michael Sala, an Australian who currently lives in Hawaii. On his webpages (exopolitics.org) on January 29, 2019 Michael Salla posted this story:

> Navy Seal Reveals Secret Mission to Ancient Buried Structure in Antarctica
>
> On January 23, Emmy award winning journalist, Linda Moulton Howe, released the video testimony of a new whistleblower discussing his highly classified mission to a large buried structure found in Antarctica. The whistleblower claims that in a classified mission conducted in 2003, he entered inside a very large octagon-shaped structure located near the Beardmore Glacier that extended down deep into the glacier's icy interior.
>
> The whistleblower is a retired U.S. Navy Seal who was first interviewed by Howe on July 19, 2018. He used the pseudonym Spartan 1 in Howe's YouTube video where his face is shadowed out and his voice is altered to protect his identity. Howe says that she personally vetted Spartan 1, who provided ample documentation to substantiate his

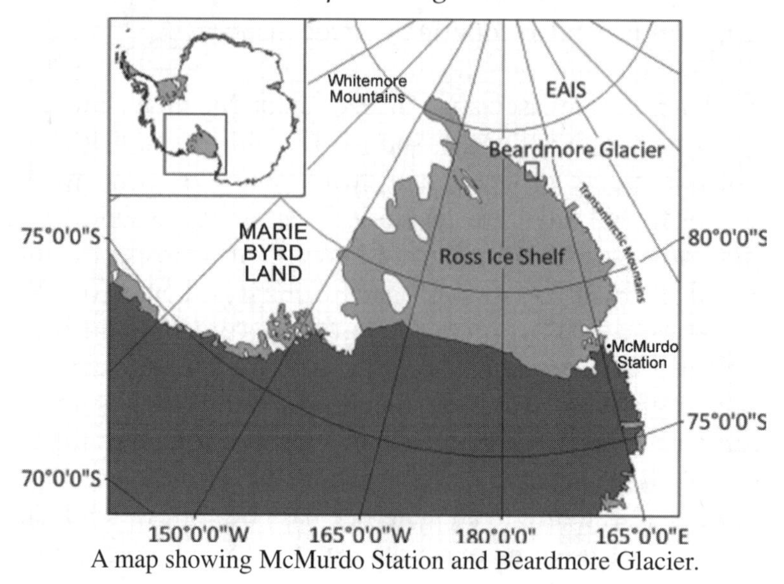

A map showing McMurdo Station and Beardmore Glacier.

military career.

Previously, Howe has released the testimony of another military whistleblower, Brian, who was a Navy flight engineer who had flown numerous support missions with the Antarctic Development Squadron from 1983 to 1997. He witnessed a number of anomalies pointing to hidden facilities or bases located deep under the Antarctic ice sheets. He says he witnessed silver flying discs over the Transantarctic Mountains, not all that far, as Howe pointed out, from where the Navy Seal had conducted his mission.

Salla then says that the Navy Seal said that ground penetrating radar had discovered the structure in the glacier and that it was an eight-sided octagon. Howe elaborated in her video:

> In 2003, a U.S. Navy Seal Special Operation team traveled to Antarctica to investigate a perfectly geometric 8-sided octagon structure discovered by ground penetrating radar near Beardmore Glacier, about 93 miles from the American McMurdo Station.
>
> Another previous team of engineers and scientists had dug out the top layer of one octagon made of a pure black substance that was built on top of two more black octagonal

structures that went down deep into the two-mile-thick ice.

Michael Salla then says on his web pages:

> In the video the Navy Seal (aka Spartan 1) described the launch of his mission from an aircraft carrier voyaging near West Antarctica's Ross Sea. He was taken by helicopter to McMurdo station, the largest U.S. base in Antarctica, and then taken by ground transport to the structure's location.
>
> Spartan 1 described entering a doorway approximately 50 feet under the ice. He estimated the walls of the structure as about 18 to 30 feet thick (6-10 meters) and the ceiling height at around 22-28 feet (7-9 meters). He said that the walls, ceiling and floor were made of a black basalt material that looked like shiny black marble.
>
> The interior was heated to around 68-72 degrees Fahrenheit (20-22 Celsius), and was also lighted by a lime green source projected from the ceiling and floor. He did not see any heating or lighting equipment, which added to the mystery of the buried structure.
>
> Only part of the structure, he stated, has been uncovered so far by the archeological teams, with the rest buried under the ice and extending far below. Ground penetrating radar had shown the structure to be an Octagon in shape, and covering an area of 62 acres (about 0.5 square kilometers).
>
> Spartan 1 described the walls and doors as being covered by hieroglyphs that were about eight inches (20 cm) high and about two inches (5 cm) deep. The hieroglyphs were neither Egyptian nor Mayan, but appeared similar to both in terms of depicting animals and other strange symbols.
>
> Significantly, one of the symbols was very similar to the Black Sun image used by the Nazi SS, who had a large version of it built on the floor of their headquarters at Wewelsburg Castle. The Black Sun image continues to be banned in Germany under their Nazi propaganda law. Howe said future episodes featuring Spartan 1 will return to the Black Sun symbol.
>
> Spartan 1 explained that part of his mission was

to transport scientists who would document the buried structure and the hieroglyphic symbols, by taking photos and making drawings. He said that his team had to leave one of the scientists behind who insisted that more time was needed to do a proper inventory of what was discovered. Spartan 1 was told and believed that the structure had been built by a human looking group of extraterrestrials, who were involved in the genetic engineering of humanity.

Spartan 1's testimony is very significant since it provides a rare eyewitness account of what is actually inside one of the buried structures whose age stretches back into antiquity. Howe's previous Antarctica eyewitness, Brian, did not actually get to see or go inside one of the artifacts. The closest he got was when he saw a large hole going deep inside the South Pole, while flying overhead through restricted airspace.

A satellite photo of McMurdo Station.

Yes, some very interesting information here, with an octagonal building of black basalt with 20-foot-thick walls and 28-foot high ceilings. There are hieroglyphs on the walls. The Black Sun symbol is seen. No bodies or skeletons are described. Presumably they would be cone-headed extraterrestrials of a tall stature.

McMurdo Station is only 93 miles away from this site, said to be on the Beardmore Glacier. McMurdo Sound is like a small town. It has all kinds of roads and vehicles and even has the occasional traffic jam. This is the busiest and most populated place in all of Antarctica and also has the most active airport.

The McMurdo Station is on the south tip of Ross Island, which is in the New Zealand–claimed Ross Dependency, however, it is operated by the United States. This station is the largest community in Antarctica, capable of supporting over 1,500 residents. All personnel and cargo going to or coming from the Amundsen–Scott South Pole Station must first pass through McMurdo Station. Nearby, and connected by road, is New Zealand's smaller Scott Base, only three kilometers away. In many ways, if Antarctica was a real country with a real population, McMurdo would be its capital.

That Spartan 1 would first land here at McMurdo makes complete sense. He says that he was transported from an aircraft carrier to McMurdo, from where he went via ground transportation to the site at the Beardmore Glacier. This seems completely reasonable. But then he says:

Some of the buildings and vehicles at McMurdo Station.

> Spartan 1 explained that part of his mission was to transport scientists who would document the buried structure and the hieroglyphic symbols, by taking photos and making drawings. He said that his team had to leave one of the scientists behind who insisted that more time was needed to do a proper inventory of what was discovered.

This does not make sense. Spartan 1 was apparently part of a party of Navy SEALs, one of only a few we can suppose, who, according to the story, were escorting scientists to a remarkable site partially frozen in a glacier in Antarctica. It is all very exciting, exactly like a Hollywood movie, but there are some big mistakes here. First of all, 93 miles from McMurdo is like a weekend snowmobile trip for Antarctic explorer-types. That a military command and control base would not have already been established at this site—by Navy SEALs themselves—is out of the question. This site would be permanently occupied immediately after it was found. What is more, personnel could easily be making daily trips most of the year to this site via helicopter (or flying disk—how about that).

So, this group arriving at an empty archaeological site in Antarctica sounds great for a Hollywood script, but not so much for reality. The same with the supposed leaving of one archeologist at the site after everyone else left. What? This is ridiculous, and is the sort of plot element that shows up in the worst of the Sci-Fi-Horror movies.

Let's face it—no scientist would be left with some boxes of supplies and a snowmobile only 93 miles from McMurdo Station to investigate this amazing structure by themselves. And in fact, as noted above, any such facility would be permanently occupied by Navy officers, scientists and specialists from the secret space program. The only way the scenario described by Spartan 1, Howe and Salla could have happened is in a movie. One I would like to see.

In the scenes after the Navy SEALs and other scientists had left the lone investigator by himself in this marvelous black basalt building with its halls and rooms—shades of the Serapeum in Egypt—the vicious long-headed aliens would stalk and kill him.

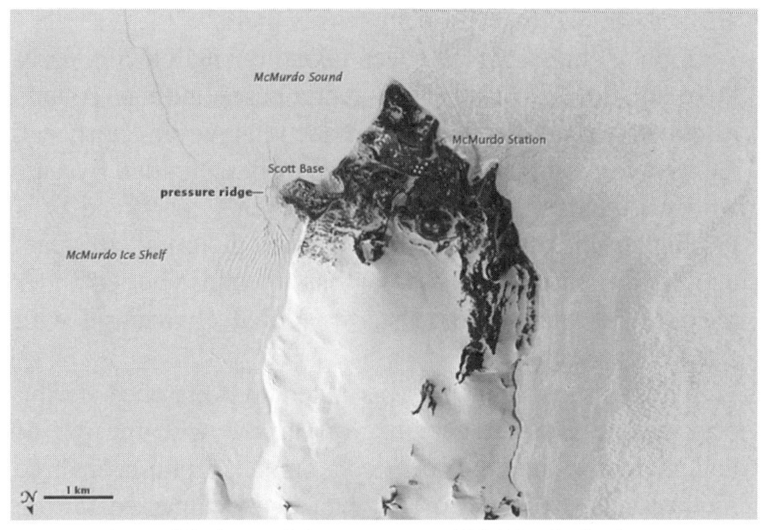

An aerial photo of the area around McMurdo Station.

Hopefully in the last act, the Navy SEALs return and save this poor guy, but he is probably already dead. They can still root out this alien nest anyway, with their flamethrowers and grenade launchers. Maybe the Arctic Men will show up too. We can already start looking forward to the sequel.

Yes, sorry, but it all seems too far-fetched to really be taken seriously, but some would like to believe things like this and I am always eager to hear such amazing Indiana Jones stories. I would love to see Navy SEALs going into such an incredible ancient structure as described, and like I always say: "Don't forget to take some photographs!" They never listen.

But let us continue with the amazing stories that have been told recently about Antarctica. Michael Salla goes on with his Internet posts and now mentions such people as Corey Goode and Pete Peterson. Both of these people have come forward and said that they have had some shocking experiences in Antarctica. Says Salla in his January 29, 2019 online post:

> To date, only two other whistleblowers/insiders have come forward to share their accounts of being taken inside or witnessing the ancient artifacts buried deep under Antarctica's ice sheets. These are Corey Goode and Pete Peterson who both say they witnessed some of the buried

artifacts during their respective visits there.

Goode claims to have been taken to Antarctica in early 2016 and 2017, where he saw secret bases and the remains of an ancient civilization buried deep below the ice sheets. He says he witnessed some bodies of human alien hybrids who were part of the genetic experiments conducted by a tall human-looking extraterrestrial race thousands of years ago. He described three very large motherships buried under the ice that were used to start a global civilization with Antarctica as a hub.

Peterson says that he was taken to Antarctica during classified missions where he was tasked with the job of understanding the advanced technologies found near three motherships, one or more of which he witnessed during his missions. Peterson's testimony corroborates Goode's account of an ancient extraterrestrial base that was used as a hub for a global civilization.

This raises some intriguing questions. Was the Black Sun symbol a pictorial representation of an ancient global civilization where the South Pole was the hub with spirals going out to its distant colonies? In the book, *Antarctica's Hidden History*, I present evidence that German nationalists using the black sun symbol established a colony in Antarctica, where they built spacecraft for deep space colonization.

According to Howe's analysis, the structure witnessed by Spartan 1 dates as far back as 33 million years, which is the general date conventional geologists give for when Antarctica was last ice free. In *Antarctica's Hidden History*, contrary evidence is presented that Antarctica was ice free as recently as 11,700 years ago, which is the approximate date for the destruction of Atlantis according to Plato. This raises the question, "how old are the ruins being currently discovered in Antarctica?"

Spartan 1's independent testimony corroborates important elements of what Goode and Peterson described, and what others claim lies hidden under the frozen continent. As more of Spartan 1's testimony is released through Howe's video series, we may get important answers to questions about what lies hidden deep below the Antarctic ice sheets.

Both Michael Salla and Linda Moulton Howe are serious researchers and have seen and heard enough to know that there are lots of strange things going on. They suspect that something is going on in Antarctica, and they are correct. However, the stories from Spartan 1 and others are very suspect. Some interesting comments concerning the so-called SEAL operative calling himself Spartan 1 are posted on webpages where these stories appear (or did) by two Navy SEALs that said:

> Comments:
>
> Avatar Chris Sims Randall Flagg • 3 months ago
>
> "He said he was a First Lt. in the Navy SEALs. I was in the Navy 13 years and I never heard of that rank IN THE NAVY. Have you?"
>
> Avatar Persona Non Grrratataaa Chris Sims • 11 days ago
>
> "That was the VERY first thing I caught as well. If you watch his interview on Ancient Aliens he actually says 1st Lt. Commander lol. Even worse. The story is TOTAL nonsense. And his grammar is awful. No SEAL is going to talk like he did. This is someone who watched too many movies. He was never "vetted" properly I can guarantee. She may have looked at some sort of military career documentation but he was never on Antarctica."

So, both of these Navy SEALs, naturally interested in activities in Antarctica, as we all are, are saying that they do not believe this story and flatly state that as far as they know, there is no First Lt. rank in the Navy. Indeed, the rank does not appear on the Navy's official web page listing ranks and insignias. So, here we have our anonymous whistleblower giving us a fake rank in the Navy for his fake story on an expedition to a fabulous basalt building inside the Beardmore Glacier of Antarctica.

That the Black Sun symbol is mentioned seems to fit into the Hollywood-Vril myth being promoted that somehow ties them into extraterrestrials. It seems unlikely, again, that the US Navy and its

Corey Goode.

special forces would leave a basalt fortress like this unoccupied, especially if the symbol of the Black Sun was found there.

Since Michael Salla has mentioned Corey Goode, this would be a good time to discuss him. I first met Corey Goode at a Whole Life Expo in Los Angeles in 2017 where we were both speaking. I was told that he was a time traveller and had been to Mars. Naturally, I was intrigued. When I told him that I knew Al Bielek, he said he was unfamiliar with him. Al Bielek was a famous survivor of the Philadelphia Experiment and claimed he time travelled and had been regressed back in time, much as Corey Goode claims. I thought that this was a bit odd for someone trained as a special military operative, in time travel, no less, to have not heard about Al Bielek.

The Philadelphia Experiment was allegedly conducted by the US Navy in 1943. It involved making the US Eldridge, docked in the Philadelphia Naval Shipyard, invisible and teleporting it, through time, to Norfolk, Virginia and another dimension. A bestselling book[58] on the subject was published in 1979 and a major motion picture followed, both titled *The Philadelphia Experiment*. Yes, apparently the US Navy was experimenting in invisibility and teleportation during WWII, while the Germans were busy with their flying saucers, cigar-shaped craft, and the mysterious "Bell." If there is a secret space program, it may well include teleportation.

Corey Goode claims that he is part of the secret space program, was specially trained as a very young man, and was ultimately sent to the Moon and Mars. He also claims that he has been in secret bases in Antarctica. His claims were at first interesting, but later they became pretty far out, as we will soon see! Other than his interviews and questions that he has answered, we don't know that much about Corey Goode since he has never written a book. Here is what Michael Sala[65] had to say about him in a July 5, 2015 interview:

> Secret Space Program whistleblower, "Corey" Goode (aka GoodETxSG), has revealed astounding details involving classified activities on Mars and the Moon. Most disturbing are his revelations about the influence of a secret NAZI breakaway civilization that successfully infiltrated the US national security system. His responses go into great detail of how secret space program activities in the US and globally, have been co-opted by unscrupulous forces and institutions that are denying humanity the benefits of the technological secrets acquired over the last century.
>
> Corey released his latest information in response to a set of email questions sent to him on May 14. In previous Q & A email sessions, public forums and his website, Cory has released details of an alliance of five extraterrestrial races called the Sphere Alliance, three of whom he has had direct physical contact with at a number of meetings involving delegates from different secret space programs. At these diplomatic meetings, he claims to have interacted not only with representatives from different secret space programs, but also representatives of different extraterrestrial civilizations, 22 of which have contributed their genetics to the evolution of humanity.
>
> The responses to the latest questions have been divided into two parts. Part one (see below) begins with Corey's

knowledge of what is happening on Mars. He claims to have personally traveled to Mars and describes witnessing a number of facilities which are owned by a space program called the Interplanetary Corporate Conglomerate (ICC). The condition of workers he saw at these ICC facilities made him suspect that they were being used as "slave labor." Surprisingly, the Corporate Conglomerate has authority over military facilities that have also been built on Mars by other space programs. Corey's response that a corporate entity essentially runs Mars using slave labor is quite disturbing.

Even more disturbing are his revelations about a secret Nazi Space Program that became operational during the Second World War despite the defeat of the Axis powers. The Nazis, according to Corey, escaped to secret bases in South America and Antarctica, where they established an alliance with a group of extraterrestrials called the Draco Reptilians. The Nazis were then able to successfully defeat a punitive military expedition by Admiral Byrd called Operation Highjump in 1947. Corey says that after a demonstration of Nazi technological superiority during the 1952 Washington UFO Flyover, both the Truman and Eisenhower administration negotiated agreements with the Nazi breakaway civilization. The Nazis then proceeded to infiltrate the U.S. national security system in ways that have undermined the independence and integrity of various US and international space programs, both civilian and military. Slave labor was a major practice in Nazi World War 2 industries, it appears that this continues with organizations that the Nazi breakaway civilization has infiltrated such as ICC operations on Mars.

The information Corey reveals is both astounding and deeply disturbing. For some it may appear too fantastic to believe. Yet, in my own due diligence of Corey's claims and credibility I have found nothing suggesting any intention to misrepresent or deceive on his part. He sincerely believes he is telling the truth about his past experiences working with different secret space programs. Other insiders have also vouched for Corey having participated in one or more secret space programs, suggesting that his claims, at the very least, deserve serious investigation.

Michael Salla had sent Corey Goode a list of 16 questions and received a reply with some rather detailed answers, which are worth looking at here. Here are some of the questions from Michael Salla and Corey Goode's answers published in full under the title, "Questions for Corey Goode on Mars, Moon and Nazi Space Program":

> Q1. You have said that the Interplanetary Corporate Conglomerate (ICC) owns most of the Mars Bases. How many Mars bases do they own, and who owns the other bases on Mars?
>
> The ICC has an entire industrial infrastructure that includes bases, stations, outposts, mining operations and facilities on Mars, various moons and spread throughout the main Asteroid Belt (where a "Super Earth Planet" once existed). They have facilities to take raw materials and turn them into usable materials to produce both complex metals and composite materials that our material sciences have not dreamt of yet. They have separate groups of facilities that produce various types of technologies as well as each facility or plant that produces a specific component of a technology so that those working in the facilities and living in the support colonies/bases do not know exactly what they are producing. Much of the time the components are multiuse and are used in cross over projects. There are facilities on Earth that operate in much the same manner that contribute to the SSP on several levels.
>
> There are other bases on Mars that are controlled by Military/Security groups as well as some scientific outposts. These can be owned and maintained by other SSP Programs but are usually going to report to the ICC on some level since the ICC controls much of the Air Space and Security Operations on and around Mars. Most of the security personnel that are assigned to Mars are assigned to and serve under the ICC. The military groups that will be

returning to their previous organizations (SSP Groups) are kept isolated from the population and personnel who live and work on the Colonies, Bases and Industrial Facilities that they protect. They are normally in the rather Spartan outposts that I have described previously in other writings. I had seen a few of these outposts built from the "Ground Up". They were always quite a distance from the main underground colonies, bases and industrial facilities and spread out in a Multi-Teared Perimeter Defensive type of system. There are "Non-Humans" also having bases on the planet. Some of them have been there for some time and have the highly coveted larger lava tube systems that have been built out into base systems that are unimaginably huge and can securely reside millions of inhabitants.

Q2. Did you ever spend time on any of the Mars bases? If so, what did you say, and how long did you spend there?

Yes, there were several occasions where there were specialty equipment malfunctions that needed to be repaired immediately. When the ICC was unable to arrange their personnel from Earth or other facilities in the Sol System to make a trip to a colony, base or industrial facility in the time needed a request was put through to the specialists on our scientific research vessel. On these rare occasions we would fly down to the location where we would be met by 4 to 6 armed guards. We would be instructed not to make eye contact or communicate with anyone for any reason unless it was directly related to the work we were there to do. In these situations there would normally be one of our security team, an Intuitive Empath, and a scientist and two technicians along with tools and parts that may be needed. We would be escorted directly to the location of the work. The local facilities security team would watch us very closely and then escort us directly back to our shuttle craft after the work had been completed and tested. We were never asked if we would like a tour, invited to spend the night or stay and share a meal with the personnel or inhabitants of the facility. We did however get a chance to see some of the people. They were usually pale, unhealthy looking both physically and mentally and seemed very much like slave labor. On more than one occasion we saw four identical people carrying crates and other items around that were obviously clones. I did notice in one colony that there was what looked like an "Art Wall" where people were hanging artwork that they had drawn and painted. This was the only time I saw anything that looked like it was meant to be positive for the metal health of the inhabitants. These were always places that we were relieved to leave. When we would visit the military outposts they were regimented but had a completely different feel or energy about them. We were also more comfortable in some of those locations because we had actually seen and been a part of them being built at an earlier date.

…

Q4. Does the Solar Warden Space Program operate any off-planet bases? If so, where are these located?

Yes, there were some stations (Located in certain areas of the Solar System where vessels would dock for repairs, conferences and personnel transfers etc…), there were also other bases on some moons and even Venus (on the surface and in the upper atmosphere). There are also some bases in nearby Star Systems that I know next to nothing about (never had a "Need to Know").

…

Q7. You have said that the Nazis established a base on the Moon, and this was later built over creating Lunar Operations Command. When did the Nazi's establish their moon base?

They had made several attempts at creating a moon base that did not go so well going back to the 1930's. They did find an ancient building that was obviously built by much larger beings that they could cement and repair enough to pressurize and use as a temporary base while they constructed the underground base that had a few visible structures on the surface one of which did take the shape of a Swastika. This base was still being built when they made the break through deals with the Americans in the early 1950's that gave them access to the Industrial Might that had cost them the European War. They now used this Industrial Power (Soon to be known as the Military Industrial Complex) to their favor and built out a massive base that went down many levels in a "Bell Shape" and the surface structures built around the former structures to become what we now call the Lunar Operation Command aka LOC.

Q8. Which group of extraterrestrials helped the Nazi's in their battle with Admiral Byrd in 1946/1947, and developing an off-world presence?

There was help from the Draco Federation as well as a group that the NAZI's were led to believe were ET's (referred to as "Arianni" or "Aryans", sometimes called "Nordics") but were actually an Ancient Earth Human Break Away Civilization that had developed a Space Program (referred to as "The Silver Fleet") and created vast bases below the Himalayan Mountains (largest in Tibet and call the system Agartha) and a few other regions. The first craft they built used Mercury turbines, and electro-gravity engines were developed by the assistance of this group. Again the NAZI's and to this day many Earth Humans who are in contact with them and others believe them to be ET's (because of their deceptions) when they are actually very much Earth Based Humans from Ancient Break Away Civilizations. I have heavily avoided speaking in depth about the few Ancient Break Away Earth Civilizations that have Space Programs and massive bases on the Earth, the Moon and elsewhere in the Sol System and other Sol Systems. Some of them have been extremely deceptive and convinced some people that they are ET groups that are here to assist Humanity. Some people have memorized some of these stories like some people memorize religious text. The Illuminati/Cabal has had a falling out with some of them and want them exposed for what they are. I think they should do the dirty work. I am not going to cause controversy that will just be a distraction from the main Blue Avian Message of Becoming Loving, Forgiving and Focusing on Expanding your Consciousness and Vibration.

…

Q15. You say that the Nazi's infiltrated the U.S. secret space program, is that the dark fleet/Cabal and/or Solar Warden, or another SSP?

Yes, I speak about it in some of the other questions here and in an article I wrote on my website. The NAZI Break Away/Secret Society Groups and their Allied ET/Ancient Break Away Civilization Groups needed the massive industrial machine of the United States that had defeated the Axis powers in World War Two. They had the Science and Technology covered but if they wanted to meet their goals of moving out into the Sol System to colonize on a large scale and setup industrial facilities and infrastructure across the Sol System to mine materials and create what they have today which has been largely achieved through the ICC Group Effort. When they had forced the hand of the United States to sign a treaty and create a joint secret space program they had already positioned operatives throughout the Military, Intelligence, Aerospace and Corporate world. They already had the Financial/Banking world in various society hands for generations before World War One. These groups were all woven in and working together with other groups that are under the Illuminati/Cabal Umbrella. The U.S. Military had known about the NAZI's landing in their exotic craft dressed up and approaching people in Europe presenting themselves as benevolent ET's from far away star systems (we know where they learned that trick) that were here to assist Humanity. After the treaty was signed and the joint Secret Space Programs began in earnest things quickly got out of hand and the NAZI Break Away group won the race to infiltrate and take over the other side. They soon controlled every aspect of the U.S. from the Financial System, The Military Industrial Complex and soon after all

> three branches of the government itself. This may be very difficult for some people to believe, however the more people are waking up to what our own Government has been up to the last 70 years the more people are coming around to realizing what has happened to the US.

Corey Goode's statements are a mishmash of popular UFO and occult material from the last few decades. As he divulges in his answers above and subsequent statements, there are apparently millions of people on Mars; some are slave labor, some are clones, some are aliens. The secret cabal has bases in other star systems and the Aryan aliens have vast ancient bases in the Himalayas. The Nazi cabal has taken over every aspect of the American government and the world economy. The Blue Avians are here to teach us love and forgiveness. He has also made claims about a "Lt. Col. Gonzales," who is in touch with this extraterrestrial group in Antarctica that are called the Blue Avians, as we will see shortly. Goode's claims of the Blue Avians are uniquely his and other so-called Solar Warden experiencers are smart to keep away from this aspect of Corey Goode's narrative.

Goode seems to prefer seeing it all as sort of a Star Wars movie, with similar costumes, badges, hats and all that. Nazis in space with some reptoid aliens.

Goode's story doesn't make sense when he says that the Nazis infiltrated every single high-tech corporation and position of power inside the now "Joint Break Away Civilization/Program," and that they have been involved in all western space programs and projects ever since. But now they occupy their own slave labor Nazi/reptoid bases on Mars, out of control of Solar Warden? Is the Dark Fleet of the Nazis different from Solar Warden? How is he saying they are the same thing but opposed to each other? Is he forgetting to tell us about the third race of reptilian extraterrestrials who have come to our solar system to take it all away from everybody? It all has the ring of someone who is good at making an intriguing narrative, but under questioning by knowledgeable people—even ones who are hanging on their every word in total belief—the stories begin to break down and reveal inconsistencies.

It doesn't matter that Goode offers no proof of his visits to

Mars, not even a photo of his Martian apartment and bathroom, but his stories seem kind of frivolous. Unlike Linda Moulton Howe's stories from anonymous informants—with ranks that other Navy SEALs have never heard of—with Corey Goode we have a first person account of visits to Antarctica, the Moon and Mars. But do they add up? Unfortunately not.

Can we believe Gary McKinnon and his revelations? Yes, I think we can. Can we believe Corey Goode and his revelations? No, probably not.

So who is Corey Goode? Is he someone who genuinely has insider information on bases on the Moon and Mars? No, it would not seem so. It would seem that Corey Goode, and all of those who promoted him, are guilty of taking a valid topic—the secret space program—and spinning it.

What seems to have gone on here is that a completely credible subject like the secret space program is suddenly hijacked (or should I say, highjumped) by obvious disinformation such as Nazi slave bases on Mars. We have yet to get to the Blue Avians in vast caverns in Antarctica, and giant alien cities beneath the ice with hieroglyphs on the walls.

In the world of secret government operations and off-world operations, people tend to disappear. Some of them are murdered for knowing too much and talking too much. Others just disappear off-world and they are never seen or heard from again.

I doubt Corey Goode will ever meet either of these fates. He is going nowhere, certainly not to the Moon or Mars. No one is going to kill him either, because he is not spilling any real secrets.

Another person who claims that they have worked for the secret space program is William Tompkins who has claimed he was part of a secret group in interviews on the Internet program Cosmic Disclosure, and he is discussed at length in Michael Sala's books.[65, 66]

William Tompkins wrote a 2015 book called *Selected By Extraterrestrials*[67] in which he claims he did secret research for the Navy, met sexy Nordic extraterrestrials with whom he has a psychic connection, and says our secret space program is fighting reptoid aliens in space. He was selected by Nordic aliens from an early age because of his psychic powers. The description on Amazon gives us a pretty good idea of what is in the book:

Bill Tompkins was embedded in the world of secrecy as a teenager, when the Navy took his personal ship models out of a Hollywood department store because they showed the classified locations of the radars and gun emplacements. He was personally present at the "Battle of L.A." when a thousand rounds of ammo were fired at UFOs, and one of the Nordic craft may have selected him to be their rep in the evolving aerospace race. This book is a partial autobiography about his life to the beginning of the 1970s including some of his early work for TRW. Selected by the Navy prior to completing high school to be authorized for research work, he regularly visited classified Naval facilities during WWII until he was discharged in 1946. After working at North American Aviation and Northrop, he was hired by Douglas Aircraft Company in 1950, and when they found out about his involvement in classified work, was given a job as a to create design solutions as a draftsman with a peripheral assignment to work in a "think tank." This work was partly controlled by the Navy personnel who used to work for James Forrestal, who was allegedly assassinated because he was going to publicly reveal what he knew about UFOs. Bill Tompkins was asked to conceive sketches of mile-long Naval interplanetary craft designs. Later, as he became involved in the conventional aspects of the Saturn Program that later became the Apollo launch vehicle, his insight to system engineering resulted in his offering some critical suggestions personally to Dr. Wernher von Braun about ensuring more reliable checkout using the missiles in their vertical position and also some very efficient launch control concepts adopted by both NASA and the Air Force. This story is peppered with very personal interactions with his co-workers and secretaries, some of whom the author believes to be Nordic aliens helping the "good guys" here on Earth. Towards the end of this volume of his autobiography, he sketches what he personally saw on TV when Armstrong was landing on the moon. Born in May 1923, Bill Tompkins is one of the few survivors of the "big war" who is still healthy, married to the same girl Mary, and is willing to tell his story about what he really did during his aerospace life in the 40s, 50s and 60s that relate to aliens, NASA and secrets that now can be told.

One review of the book on Amazon said:

> A laughably badly written piece, repetitive and childish, it consists of much timeline shifting, patently bald inaccuracies and inane conversations of supposed accomplishments written in the first person narrative of someone conversing with peers, quoting such incomprehensible statements that it's difficult to have a break from reading despite the awful writing.
>
> If Mr. Tompkins was as highly qualified and held such esteemed positions in such secretive environments, it goes against the grain that he would have been allowed to make even a fraction of it public. The subject matter is fascinating however and 99% so ridiculously unbelievable and ridiculous that it is probably all true, you'll have to be 15,000 years old and be a Reptilian from beyond Zeta Reticuli to appreciate or understand most of it. Don't take it too seriously and just enjoy it for what it is but if you are after riveting story telling stick to Asimov, at least you know that's fiction.

Another review said:

> Got this in a moment of delightful dizziness. As others have noted, this is a load of repetitive rambling. He's only just introduced one of his alien porn stars but I can't take any more of the disjointed musings. He keeps on going on about the aliens being thousands, if not millions, of years ahead of us. Why oh why would they need us to build ships to support them in their interstellar/intergalactic wars? Look how far our own AI and robots and so on have come. It would be much easier for the aliens to make 'droids rather than have us build support-fleets—pffft. It makes absolutely

no sense at all. I'm not saying I don't believe in black ops having and working on anti-gravity technology—of course, they are. But I still have a couple of questions regarding alien motivation:

1. What are they actually here for? I ask because mineral resources are much more plentiful and easy to get in the asteroid belt. Even for genetic stock makes no sense for aliens who can do genetic engineering on top of faster than light interstellar travel.

2. Since it would be far easier to knock up a couple of robots, why would they genetically engineer humans, who are weak and not robust, as slaves?

...I'm more than happy with the idea of there being a number of alien tourists visiting Earth for whatever reasons. However, I can't imagine them being so technologically balanced that they are fighting it out here. I imagine there would be some kind of overarching inter-alien organization. Anyhow, whichever way you look at it you'd think that if somebody was telepathically communicating with aliens (who are fighting interstellar battles) why wouldn't they at least let him know what their motivations are?

The review continues, making several good points about the improbability of Tompkins' accounts. Indeed, Tompkins (with the help of his sexy Nordic alien secretary) seems to be taking his own UFO research from the 1960s, 1970s, 1980s and onward much too seriously, and has decided to put all of his beliefs and fantasies into a difficult to read book, complete with manufactured and distracting conversations that he supposedly had over the years with important people (and aliens).

A similar book is *Genesis for the Space Race: The Inner Earth and the Extraterrestrials*[68] by John B. Leith. Says the description for the book on Amazon:

John B. Leith was born in 1920 with a photographic mind. He graduated from high school in 1937 and took spiritual training in Lhasa, Tibet for nine months before joining the Office of Strategic Services. During the last two years of

WW II he led a six-man special-forces team out of London specializing in breaking prisoners out of German jails in France, Germany, and Poland. Though he became the most highly decorated soldier of any war, he was not allowed to acknowledge he had any combat service until after his death. He studied literature at Oxford while describing his 12 most exciting of 62 missions behind German lines. As liaison officer between General Patton and Gen Von Rundstedt he learned how selected people were fleeing Germany. These missions are described in *The Man with the Golden Sword.* Leith studied in Canada and spent 25 years earning a living as a journalist. In 1975, after demonstrating that he could keep his OSS experiences hidden from the public for 30 years, the CIA asked Leith to team up with another wartime OSS officer, Father John in WW II, and interview people associated with the development of "round-wing planes," both for our allies and for Germany. Neither group was allowed to use that technology for war, because both were assisted by wiser beings from off this planet. *Genesis for the Space Race: The Inner Earth and the Extraterrestrials* was dedicated to "an unsung hero, Jonathan E. Caldwell, inventor of this civilization's first spacecraft." He used several names for security. Read about him on Wikipedia. Like most CIA-approved releases, this 376-page book was required to have disinformation published with it and shown in a map of inner-Earth civilizations provided by an Atlantean cartographer. A 4-page Preamble helps the reader understand the disinformation and it's need. This book falsely indicates Hitler went with other officers and scientists to New Berlin, accessed from Antarctica. Pope Pius XII trained him to convert pagans in the jungles of Ecuador. The map falsely shows 125-mile-diameter holes at each poll required by the CIA to keep wealthy people from spending their fortunes in attempts to access the Inner Earth to acquire advanced technologies. Leith's term "Inner Earth" was sometimes changed to "Hollow Earth" and the story modified to match. Also, this book says that Hitler's son came to his father's funeral in Zaragoza, Spain in 1974,

> though the real funeral to which his son, born to Eva Braun in 1945, came was in Cuenca, Ecuador in 1978. It has three sections: Space Race, The Inner World of Extra Terrestrials, and Primer For a New Age of Space. Chapter XVII is titled Intrusion of Alien Beings into World Societies and Interviews with Leading Extra Terrestrials. The Epilogue describes how the 54-12 committee developed strategy for USA presidents since Eisenhower to maintain air supremacy into the 1980s and beyond. It describes foreign awards presented to Lt. General Caldwell after he retired in 1967, including a singular Congressional Medal presented by Eisenhower. The Appendix is titled Social, Political, Economic and Religious life in Inner Earth. Notes and Sources follow with excellent research on many controversial issues that came up during the interviews. Following are pictures of construction of New Berlin and a completed housing block with many German soldiers lined up for inspection.

I was shown this book at a conference a few years ago and went to an online bookstore and bought a copy. Leith actually died in the year 1998 and his manuscripts were published years later by his friends Robyn Andrews and Donald Ware. *Genesis for the Space Age* was published in 2014 and does have a curious photo and document section in the back with some maps of the inner earth showing the location of such inner earth cities as New Berlin, New Hamburg, Memel and Himmler. These cities, with large populations, are shown on the map to be connected by railways. There are islands and seas in the inner earth which also have cities and railway lines. One of the large islands is called Agahrta and has such cities on it as Zennud and Shambala (yes, they are connected by trains). This map was drawn by Robyn Andrews.

There are some old photos too, mainly of scenes from WWII, of Admiral Byrd, and such, but there are a few curious photocopies of photos that purport to show the construction of huge buildings inside the earth at New Berlin, a city of over a million people. The photos do seem to show large construction projects, including some railway lines, but there is no way to know that they are inside the earth; and they could be poor photocopies of construction from practically anywhere in the world. There are no signs and symbols, such as a swastika or Black Sun, only a bunch of concrete and steel bar construction.

Leith's book, like William Tompkins', is really a first person story full of fabricated conversations with people that stumble the narrative forward so that Leith can discuss everything from secret OSS operations during the war (the OSS was the American Secret Service until the creation of the CIA after the war, but Leith seems to be a Canadian, so how he came to work for the American OSS is never really explained) to Operation Highjump and Operation Paperclip to secret bases in Antarctica, Admiral Byrd flying into the hollow earth and a whole Nazi world inside the planet. No need to recap all of the hollow earth lore here as I did in chapter four, but let me say that Leith includes all of it in his book and then some. The bizarre map hand-drawn by Robyn Andrews is also full of just about every reference to the hollow earth that there is. It is the one unique illustration in the book.

Unfortunately, Leith's conversations throughout the book seem forced and fake, like in William Tompkins, and ultimately they add to the fiction of the whole book, rather than making it look like it is some true story and confirming that Leith's bizarre life was real in some way. The book rather shows the Leith led an interesting life and was keenly interested in Tibet, Agartha, Shambala, the Nazi SS, the hollow earth, UFOs and more. He, like William Tompkins, is a psychic and much of what he knows apparently comes to him psychically. Both of these books should be considered fiction presented as fact.

In a June 19, 2016 blog (https://divinecosmos.com/davids-blog) David Wilcock goes on at length about a literal UFO war going on around the planet, somehow missed by most of us, and then goes on to promote Corey Goode and his claims as if his revelations are what his readers have finally been waiting for. Goode even provided Wilcock with a map of Antarctica with five spots where the underground bases of the cabal were located. Wilcock says:

> There are a series of large, habitable regions underneath the ice of Antarctica that are host to certain groups, including those of the Cabal. This works much the same way as an

Corey Goode's map indicating the six secret bases in Antarctica he identified.

igloo in Alaska, where even though the walls are made out of ice, a small fire inside can keep it nice and warm. Natural ice caves are created by the heat of sub-surface volcanic activity. There is land you can walk and build on, breathable air, and running water. Here is an image commissioned by Corey Goode of one of the Cabal's own bases underneath the Antarctic ice, complete with buildings and exotic craft.

Corey Goode recently provided us with a map of six different naturally volcanic areas under Antarctica that are home to secret Cabal bases. Each red circle corresponds to an area that has been developed and has people living there comfortably right now. These areas also show up on heat maps of Antarctica as visible volcanic hot-spots.

Access to Antarctica is strictly controlled, meaning we cannot just go and check it out for ourselves. The freezing cold of the surface area also makes it very difficult for any individual or group from our ordinary world to try to go sightseeing. According to the 94-year-old aerospace engineer William Tompkins, there are also two much larger habitable caverns off to the right of Antarctica.

Their combined landmass is quite significant, dwarfing the size of either of the largest two ovals on this map. They are almost vertical, on a slight diagonal angle, and run from the top of the continent to the bottom—creating a vast amount of livable space.

DRACO BASES IN ANTARCTICA

Both of these enormous strips of land are under the exclusive control of reptilian ETs known as the Draco. According to Pete Peterson, and confirmed by Goode and others, the Cabal has often referred to this type of ET as Saurians.

They have vast, densely populated cities in these areas, composed of buildings that are far more high-tech than anything we see on the surface. The Draco are on the losing end of a war througout the galaxy that has beaten them back into our solar system and a few neighboring ones. The Draco have built vast cities in these two caverns below the Antarctic ice, with populations numbering at least in the millions.

Their total population numbers in our solar system could now be as many as seven billion — comparable to our own native population on Earth. We are protected from them by benevolent ETs that will never allow them to do more than we have invited by our own collective free will. Furthermore, their time here on Earth is very likely about to come to an end.

TOMPKINS VALIDATES INTEL

According to Tompkins, the Draco cut a deal with the Germans before World War II, and offered the smaller areas in the above image as a concession prize. This allowed the Germans to begin developing their own secret space program, using technology given directly to them by the Draco. The plan was to create a "Dark Fleet" of ships and personnel that could go out and begin conquering other worlds, as the Nazis attempted to do here. The Draco would be leading the effort, and needed the industrial might of the Germans to help produce a sufficiently large army to make it possible.

THE 'ANSHAR' HAVE A SIGNIFICANTLY ALTERED PERCEPTION OF TIME

The white craft you see in the middle of the under-ice cave image is what we are calling a "bus," built by another humanlike group living inside the earth. This group, who refer to themselves as the Anshar, have a much more accelerated perception of time than we do.

Thus, events that took place thousands of years ago, as we measure time, could be considered as fairly recent history for them. Almost all of the things we cover on *Ancient Aliens*, including giant stone monuments and even Atlantis, are considered very recent by these standards. The arrival of Jesus, and the angelic beings that helped guide us through those interesting times, was also a last-minute development from this vantage point. This gives us a remarkably different perspective on how we see such civilizations.

The Anshar only came into the scene as of last September, but they have now become one of the most interesting and important elements of our story.

They routinely maintain telepathic contact with many of us, usually subconsciously. I have had clear and obvious messages from them as well.

They do appear to have a key role in helping us go through this epic transition that is heading our way.

THE ARRIVAL OF GIANT SPHERES

Pete Peterson revealed that a Neptune-sized object called "The Seeker" came into our solar system in the 1980s, terrifying the Reagan Administration. Many more spheres, usually about the size of the Moon, appeared in the late 1990s and early 2000s.

They came in, cloaked themselves, took a position and then just remained there. They did not answer any hailing signals. There were plenty of visible signs of these arrivals in the "Sun Cruiser" phenomena many observers were catching off of NASA's SOHO satellite photos. At least 100 more spheres arrived around the time of the Mayan Calendar end-date of December 21, 2012.

Now, they could be as large as Neptune or Jupiter in size. Again they assumed various positions and then simply remained silent and motionless. …Corey and I presented many visible and stunning images of these planet-sized objects in our public debut at the Conscious Life Expo this

> past February.
>
> …Corey was taken up there by a very unusual means, involving a small orb of blue light that turned into a sphere that engulfed him and brought him into space.
>
> THE BLUE AVIANS AND LT. COL. GONZALES
>
> In this initial, landmark meeting, Corey also was re-acquainted with ETs he had experienced contact from before, known as the Blue Avians. His initial contacts with these beings were so sacred to him that he refused to tell me anything about the experience. Other so-called "sphere beings" were introduced to him, and to the Alliance, once he was brought up in this same meeting. Corey's main human contact with the Alliance was through a man we are calling Lt. Col. Gonzales, who had also been approached by the Blue Avians.
>
> Gonzales had the very unique job of liaising between the SSP Alliance and other groups working to defeat the Cabal that we are calling the Earth Alliance. A coordinated effort is underway to produce Full Disclosure, where we get all the technology and benefits the SSP now enjoys. It has been awkward for both of us, to say the least, to be given such an outrageous-sounding story to share with the public. However, the daily realities this situation has thrown us into are as serious as a heart attack, requiring constant vigilance to remain safe and effective.

The blog goes on, but I will stop here. It is all pretty alarming stuff, this battle going on in space, the terrible cabal that is keeping us all under control, the secret underground caverns in Antarctica full of Dracos, Blue Avians, Nazis and more. We have "Lt. Col. Gonzales," who can vouch for Corey Goode if he ever comes forward. Fortunately the Anshar have just shown up (2015) and are in telepathic contact with light seekers who want to free our planet. David Wilcock himself has had telepathic communications from them (see above: "I have had clear and obvious messages from them as well."), and so he knows that Corey Goode is telling the truth, as are the other people he mentions in his blog.

It is impossible to know what David Wilcock thinks of Corey

Goode these days, but it does not give any of them credibility to espouse these stories and, if anything, it makes serious researchers wonder if these people are just self-centered gullible believers or if they actually have some sort of disinformation agenda. The one thing that all of these people do, including Michael Sala, is make serious readers discount about everything they say and "throw the baby out with the bathwater." In other words, they make the idea of any real UFO activity in Antarctica seem foolish and the discussion of a secret space program one of outlandish stories and space invaders of all different types. Goode continues to promote himself and after years of drifting between UFO conferences now has his own website. He continues to promote the Blue Avians (can't stop now…) whom he says are avian humanoids with flexible beaks. They speak telepathically and, at the same time, make hand signals using one hand. The Blue Avians will send a blue sphere to pick up contactees and transport them to a giant blue sphere. The floor takes form, which is a series of crisscrossing lines of white light in a square grid pattern. On Goode's website (https://www.spherebeingalliance.com) we learn that he is a Blue Avian himself and that he is the reincarnation of the Prophet Enoch:

> A member of the Blue Avian soul group who incarnated to Earth to help with planetary ascension, Corey was recruited into the Secret Space Program at the age of six because of his intuitive empath abilities.
>
> Since 2015, Corey has been receiving messages from Blue Avian, Ra-Tear-Eir, and from Ka Aree from the inner Earth civilization known as the Anshar. The Blue Avians refer to Corey as "Ra-Hanush-Eir"—a derivative of a Hebrew word which is synonymous with "Enoch" and roughly translates to "Messenger." Quite literally, Corey is the Enoch of our modern times—sent to our planet to reignite the Christ Consciousness message of love, forgiveness, and service to others in preparation for the most extraordinary time in our recorded history.
>
> The message of the Blue Avians is that humanity needs to be more loving, forgiving and in service to other selves. The Anshar are our distant future relatives, who have come

> back in time to manage their timeline and ensure their (and our) survival. Our beloved planet Earth (Gaia) is currently moving into a part of our galaxy that is being bombarded with highly-energetic cosmic waves.
>
> This energy is the impetus of our planetary ascension; and those whose frequency is not a vibrational match will find the ascension process very difficult.
>
> Corey's class will discuss what Corey has learned from the Blue Avians and Anshar about the ascension process, and how to raise your vibrational frequency so that you can experience ascension in harmony with Gaia, as our Creator intended.

Goode seems comfortable creating his own special extraterrestrial reality, even to the point of inventing new extraterrestrial groups, since the Dracos, Reptoids, Greys and Nordics are not enough for his universe. He is now the Prophet Enoch as well and is kept in constant touch with the Anshar extraterrestrials of the Inner Earth. Fortunately, David Wilcock is in contact with them telepathically as well and can verify much of the material. Yikes!

But, if we can discount all of the far-fetched accounts from Corey Goode and his friends, what are we to make of the real mysteries of Antarctica? Is there a secret space program like Solar Warden? Did the Nazis have secret bases in Antarctica after the war? Is something strange going on in Antarctica today? I think that we can basically say yes to all these questions. There is a secret space program and Antarctica does have some involvement with it. What is at the South Pole anyway, a research station or a secret space base?

The Crucial Base at the South Pole

At the South Pole is the Amundsen–Scott South Pole Station, officially a United States scientific research station. It is located at the geographic South Pole, on the polar plateau, at an elevation of 9,300 feet (2,835 meters) above sea level. The station sits on an ice sheet that is over a mile thick at 8,858 feet (2,700 meters) thick. The polar base drifts with the ice at about 33 feet (10 meters) a year toward the Weddell Sea. Just to be clear: there is no base of any kind at the North Pole, but at the South Pole there is an airport and a

The R4D-5L Navy plane that landed at the South Pole in 1956.

small town. Parts of this station are beneath the ice. Many people are completely unaware of this and we will discuss this matter again shortly.

Since the Amundsen–Scott Station is located at the South Pole, it is the only inhabited place on the land surface of the Earth from which the Sun is continuously visible for six months, and which is then continuously dark for the next six months. Thus, during each year, this station experiences one extremely long "day" and one equally long "night." During the six-month "day," the angle of elevation of the Sun above the horizon varies continuously. Even in the summer, the temperature is always below freezing with a typical summer high of –28 °C (–18 °F). During the six-month "night," air temperatures can drop below −73 °C (−99 °F). This is also the time of the year when blizzards, often with gale-force winds, strike the Amundsen–Scott Station. Despite these blizzards, the continuous period of darkness and dry atmosphere make the station an excellent place from which to make astronomical observations.

As mentioned earlier, the original station was built in 1956-1957 and is now buried beneath the ice. Incredibly, the station has been continuously occupied since it was built and has been rebuilt, demolished, expanded, and upgraded several times since 1956. The station was moved in 1975 to the newly constructed Buckminster Fuller geodesic dome 50 meters (160 feet) wide by 16 meters (52 feet) high, with 14 meter (46 feet) high steel archways. One archway served as the entry to the dome and it had a transverse arch that contained modular buildings for the station's maintenance, fuel bladders, power plant, snow melter, equipment and vehicles. Individual buildings within the dome contained the dorms, galley, rec, post office and labs for monitoring the upper and lower atmosphere and numerous other complex projects in astronomy and astrophysics. The station also included the Skylab, a box-shaped tower slightly taller than the dome. Skylab was connected to the Dome by a tunnel under the ice. The Skylab housed atmospheric sensor equipment and later a music room. The Dome was dismantled in 2009-10 and removed from the continent. The current station was dedicated on January 12, 2008.

Most Antarctic Program personnel and cargo reach the South Pole from McMurdo Station via ski-equipped aircraft, whereas most of the fuel used at the South Pole station is transported via surface traverse from McMurdo Station. The short austral summer, when most activity occurs, is from late October through mid-February. The station is isolated for the rest of the year and only a few personnel are in residence. The winter population is around 45, and the summer population averages 150.

Research at the South Pole officially includes astronomy, astrophysics, aeronomy, auroral and geospace studies, meteorology, geomagnetism, seismology, earth-tide measurements, and glaciology. One might think that they study other things as well, including UFOs.

It is interesting in this context to revisit the information received by Linda Moulton Howe, from her informant, Brian. He said his unit was was not allowed to fly over a certain area five miles from the station because it was an "air sampling camp." They did, however, on two different occasions, fly over this area and saw a huge hole in the ice. According to Brian, "You could fly one of our LC130 into this thing." There was talk about a UFO base at the South Pole and speculation that this hole had something to do with it.

So, are we looking at a new incarnation at the famous "hole at the pole" scenario with hollow earth believers? After all, it has now been pretty well established that instead of a portal into the hollow earth at the South Pole there is an airport. Maybe it is a spaceport, rather than an airport. Or, is the spaceport five miles away from the Amundsen-Scott South Pole Station? Google maps has supposedly

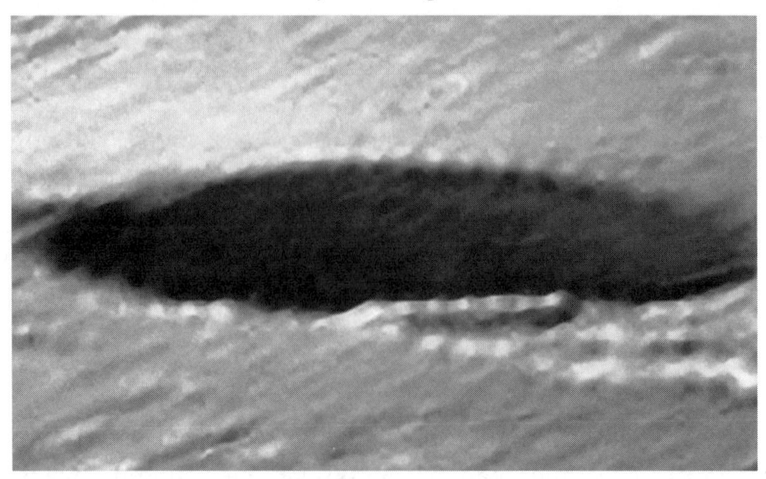

A Google Earth photo of what is apparently a large hole in Antarctic ice.

produced a photo of what appears to be a huge hole in the ice. Is it the area five miles away from the pole that "Brian" was talking about? Is this some secret under-ice base that has the latest in Solar Warden earth-based flying saucer craft? Certainly, this is intriguing, and we need to remember, the ice here is more than a mile thick.

The idea that UFO work is going on at the South Pole is interesting, and that some of the craft are extraterrestrial and that extraterrestrials established themselves around the pole are the premises behind the story promoted by David Wilcock and others as to why Buzz Aldrin had gone to the South Pole in December of 2016. Supposedly Buzz Aldrin flew to the South Pole base from South Africa, was shown alien artifacts and had a heart attack. He was then flown to New Zealand where he was treated by a doctor named David Bowie.

There is something a little strange about Aldrin's trip to the pole, where he did in fact have a heart attack. According to a *Washington Post* story, Aldrin, who in 1969 became the second person to walk on the moon, arrived at the South Pole on November 29, having flown there on a plane with the South African tour company known as White Desert. According to a message from Aldrin's assistant, Christina Korp, he was scheduled to be there until December 8, but his "condition deteriorated" shortly after arriving. Normally tourists who fly to the South Pole with White Desert stay for only one night at the pole, if that, but Aldrin was staying for more than a week.

Aldrin was evacuated to McMurdo Station and then taken to

a hospital in Christchurch, New Zealand, where he spent a week recovering from congestion in his lungs. His manager described the evacuation as "grueling" at the time, but noted that Aldrin was in good spirits as his body responded well to antibiotics.

In a statement posted on Twitter while he recovered in the hospital, Aldrin noted that his trip to the South Pole followed other "exploration achievements," such as spacewalking "in orbit during the Gemini 12 mission in 1966," walking on the moon and traveling to "the Titanic in 1996 and to the North Pole in 1998." The *Post* quoted some of Aldrin's comments:

> I'm extremely grateful to the National Science Foundation (NSF) for their swift response and help in evacuating me from the Amundsen-Scott Science Station to McMurdo Station and on to New Zealand," the statement said. "I had been having a great time with the group at White Desert's camp before we ventured further south. I really enjoyed the time I spent talking with the Science Station's staff too.

White Desert specializes in trips to Antarctica from Cape Town, South Africa. They currently offer a trip to the South Pole station for around $100,000 dollars. This trip involves a five-hour flight from Cape Town on a Gulfstream G550 private jet. This G550 lands at a private camp in the part of Antarctica nearest to South Africa, called Wolf's Fang Runway. The plane then flies to a nearby base called Whichaway Skiway. For the trip to the South Pole a different plane, a Basler BT-67, takes off from Whichaway Skiway and travels to a site called FD-83 where the plane is refueled. It then departs for the South Pole, where it lands at the Amundsen-Scott Station, a two-hour flight from this refueling station. The flight to the pole from Whichaway Station is total of seven hours.

The group then stays at the station for a few hours or perhaps one night. The small group of tourists get their passports stamped at the South Pole, take a bunch of photos and then depart. The White Desert brochure says that, depending on a variety of factors, you might spend the night in specialist tents at the pole or fly back to FD-83. On the next day the polar tourists fly back to Whichaway

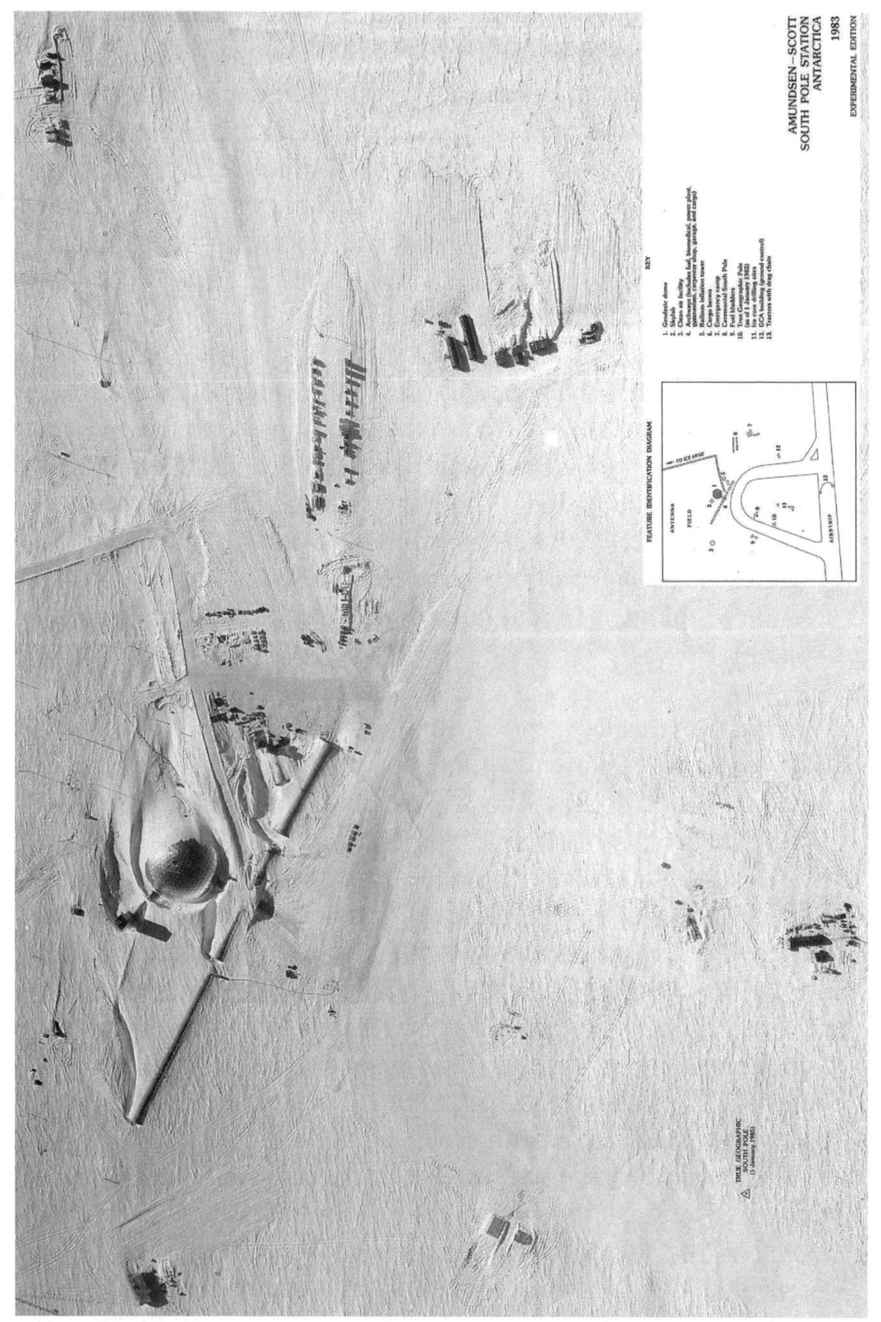

An aerial photograph of the Amundsen-Scott South Pole base from 1983.

Skiway, where they spend the night again, and then ultimately return to Cape Town.

In Buzz Aldrin's case, he flew to the South Pole, had a heart attack, and ended up being flown to McMurdo Station in an emergency flight and from there to New Zealand where he received hospital care. But, he was scheduled to stay for a week. What was he planning to do? Ostensibly, he is interested in the frozen continent because it simulates the environment on Mars, where he has been a big proponent of going.

Aldrin was reported to have tweeted on his own account: "We are in Danger, it is Evil Itself." Did he really do this? What did he mean by this? Was this an Internet hoax?

The news site Snopes.com gives us the full story posted on December 13, 2016 under the title: "Buzz Aldrin Tweeted 'We are all in Danger, it is Evil Itself'?" The Snopes.com article said:

> On 7 December 2016, conspiracy web site SuperStation95 published a post reporting that 86-year-old astronaut Buzz Aldrin (who had just been evacuated from the South Pole due to health concerns) had mysteriously and alarmingly tweeted a photograph of a mountain along with a chilling comment:
>
> His words: "We are all in danger. It is evil itself," and showed a photograph of a pyramid located at the South Pole as shown here.
>
> Mr. Aldrin seems to be referring to this object, as seen in a serial photo from Google Earth… As has been reported by worldwide media, US Secretary of State John Kerry recently traveled to the South Pole, allegedly to become better informed about "Climate Change." Kerry was the highest-ranking US government official ever to visit the South Pole, and his visit struck many as unusual. Few saw any purpose whatsoever to sending America's top Diplomat to the farthest reaches of the earth to see … ice.
>
> Now, with Buzz Aldrin's tweet, and its strange deletion, folks are wondering if sending America's top Diplomat to no-man's land, perhaps had something to do with Diplomacy after all. Is there some entity there with which we need

Diplomatic contact? …If so, why did Buzz Aldrin warn that we are all in danger? Why did he call it "Evil itself?"

The first tweet did not have a date, share count, number of replies, number of favorites, or number of retweets; all could be used to determine whether Twitter's archive backed the claims made by the sites. Additional irregularities existed in the background display of Aldrin's account: The suspect tweet didn't display in the same manner it did in our screenshot. Aldrin's account data was oddly arranged, and his blue verification checkmark was missing.

SuperStation95 did not provide a date, time, or sturdy source for the screenshot of the tweet. Using Twitter's "advanced search" tool, we were able to determine that no retweets or "modified tweets" (a form of copying and pasting a retweet) existed anywhere on Twitter. The first verbal mention of the purported tweet appeared on 8 December 2016, one day after the blog post linked above.

It wasn't just skeptics questioning the authenticity of the claim. YouTube's SecureTeam (the self-described "number one source for breaking news and exposure of the alien phenomenon") examined the claim, quickly dismissing it as a forgery.

After appearing in late 2015, Superstation95 began using what were often legitimate tragedies or unfortunate events (such as Buzz Aldrin's evacuation from Antarctica in December 2016) with falsified and often frightening details. Among the most prominent were claims that a large group of Muslim men had fired randomly upon campers and hikers in California, Fukushima radiation caused severe mutations in ocean-dwelling creatures, cargo ships mysteriously ground to a halt signaling nothing short of an imminent global economic catastrophe, a deadly Las Vegas strip car crash involved a driver shouting "Allahu Akbar," the San Bernardino shooting was provoked by pork served at a holiday party shortly before the massacre, the Earth's "magnetosphere" inexplicably collapsed for two hours, a (nonexistent) suicide note left by a genuinely deceased ICE agent revealed impending FEMA camp implementation

> and mass enslavement, and that articles about the Orlando shooting appeared on Google News hours before the attack. Although the pieces sometimes contained details of real-life events, the claims were always willfully distorted to peddle a false and upsetting conclusion. Superstation95's opportunistic leveraging of tragedy has never borne out as the real thing, despite the innumerable grandiose claims made since the site's inception. However, despite them, no one has yet gone to FEMA camp, the global economy has not melted down, the magnetosphere caused no space-based catastrophes, and no hikers were reported injured or dead as a result of roving Muslim shooters.

It should be noted that Superstation95 no longer exists as a website and their cutting edge "fake news" is no longer being put out to the eager masses. It is interesting that Buzz Aldrin went to the South Pole, but things can get blown out of proportion.

But this does not mean that there is not a secret space program, nor does it mean that Antarctica is not involved. It seems almost assuredly so. When we look at the early German expeditions and the things that went on after WWII we can only assume that Antarctica was an area of intense UFO activity for many decades. Most people, even those interested in Antarctica, are unaware of an occupied base at the South Pole, its airstrip and other facilities.

In 1991 the famous British comedian and television personality Michael Palin did a BBC television documentary called *Pole to Pole*. In this series Palin flies from country to country starting at the North Pole and eventually ends up at the South Pole. Michael Palin visited the base on the eighth and final episode.

The trip from the North Pole to the South Pole went via Scandinavia, the Soviet Union, parts of Europe, and through the heart of Africa. The intention was to follow the 30 degree east line of longitude, which would cover the most land. He had originally planned to go to Antarctica from South Africa, but a last-minute diversion to Chile included South America in the series. Using aircraft as little as possible, the whole trip lasted just under six months.

In the last episode of the series a travel adventure company is able to take Palin to the South Pole from its base in Chile. This means Palin must abandon the 30 degrees east meridian. Early in the episode he travels by airplane from Cape Town to Rio de Janeiro and then to Santiago, Chile. After having lunch at a fish market with his guide while listening to a pan flute player, he is off to Punta Arenas in southern Chile.

From there he waits anxiously for the weather conditions to allow the trip to Antarctica. After a couple of days, Palin and his crew fly on a 1953-built Douglas DC-6 plane to a base camp at Patriot Hills in Antarctica. While there, he again has to wait for the go-ahead to set off for the South Pole. Finally, after a day, he makes a final flight to the Amundsen–Scott South Pole Station but they radio him not to come and to turn back. He lands at the South Pole anyway on December 4, 1991. He visits the spot marked as the South Pole, becoming one of the very few people who have visited both the North and South Poles. He ventures into one of the buildings at the Amundsen–Scott South Pole Station where he learns that the base is occupied by American and Russian military personnel and goes for 12 stories beneath the ice. He is not allowed to see this part of the under-ice base. He then departs in his DC-6

The dome and a portion of the other buildings at the South Pole base.

back to the Chilean base at Patriot Hills and back to Chile.

What was remarkable about Palin's program when it first aired in Britain and the United States was that most people were completely unaware that a joint American-Russian base was at the South Pole. Nor did people realize how extensive this research station was, that it was permanently occupied, and the airspace above the pole was a "controlled airspace." Why was that? What were Russian personnel doing at an American base at the South Pole? Wasn't Russia our enemy? Why were we cooperating with them?

The Russians have their own base relatively close to the South Pole Station; it is the Vostok Research Station, founded in 1957. It is about 808 miles (1,301 kilometers) from the South Pole, in the middle of the East Antarctic Ice Sheet. The Vostok station is one of the most isolated established research stations on the continent and when occupied is supplied from the Russian Mirny Station on the Antarctic coast. The station typically has about 25 scientists and engineers during the summer months.

In the 1970s the Soviet Union drilled a set of cores 500–952 meters (1,640–3,123 feet) deep. These have been used to study the oxygen isotope composition of the ice, which showed that ice of the last glacial period was present below about 400 meters' depth.

Lake Vostok, beneath the Vostok Station, is 250 kilometers (160 miles) long by 50 km (30 miles) wide at its widest point and has an average depth of 432 meters (1,417 feet). The lake has an estimated volume of 1,300 cubic miles (5,400 km3), making it the sixth largest lake in the world by volume.

The lake is divided into two deep basins by a ridge. The liquid water depth over the ridge is about 200 meters (700 feet), compared to roughly 400 m (1,300 feet) deep in the northern basin and 800 m (2,600 feet) deep in the southern. The lake is named after Vostok Station, which in turn is named after the *Vostok*, a sloop-of-war ship, which means "East" in Russian.

The existence of a subglacial lake in the Vostok region was first suggested by Russian geographer Andrey Kapitsa based on seismic soundings made during the Soviet Antarctic Expeditions in 1959 and 1964 to measure the thickness of the ice sheet. The continued research by Russian and British scientists led to the final confirmation of the existence of the lake by 1993, and drilling into

A photo of Russia's Vostok Station.

the ice began. After several stops and starts, the ancient Lake Vostok was finally breached on February 5, 2012 when scientists stopped drilling at the depth of 3,770 meters and reached the surface of the sub-glacial lake. There was plenty of speculation that they would find ancient microbes in the lake, even extraterrestrial ones.

Bizarrely, the Russian RT News (rt.com/news) ran this story two days after their drilling had successfully reached the lake: "Lake Vostok mystery: Alien life, global warming and Hitler's archive" (February 7, 2012). The story then read:

> Scientists, environmentalists and even World War II historians have reacted with a mixture of excitement and concern to news that Russian geologists have drilled through to a huge subglacial lake in Antarctica, some 20 million years old.
>
> It has taken more than 30 years to work through 3,700 meters of thick ice – drilling in temperatures as low as minus 80 centigrade. But it will have been worth it, if even half the claims being made about the lake are true.
>
> **Life not on Earth**
>
> Sealed off below the ice for millions of years, the lake is a unique environment.
>
> "According to our research, the quantity of oxygen there

exceeds that on other parts of our planet by 10 to 20 times. Any life forms that we find are likely to be unique on Earth," says Sergey Bulat, the Chief Scientist of Russia's Antarctic Expedition to *Russian Reporter* magazine. "But there is one place not on Earth that has similar conditions – Europa, the mysterious satellite of Jupiter."

"The discovery of microorganisms in Lake Vostok may mean that, perhaps, the first meeting with extra-terrestrial life could happen on Europa," said Dr Vladimir Kotlyakov, Director of the Geography Institute at the Russian Academy of Sciences to *Vzglyad* newspaper.

So far scientists disagree about the presence of life forms in the water. A lot depends on how the lake was formed. If the lake formed when Antarctica was already frozen—as ice was melted by the Earth's core—then the chance of finding interesting micro-organisms are slim. But if the lake already existed when the Antarctic was still warm, anything is possible. Non-scientists have asked if these life-forms could be dangerous—undiscovered viruses, or perhaps even a monster, like that in the John Carpenter film *The Thing*. "Everything but the samples themselves will be carefully decontaminated using radiation. There is no need to worry," Valeriy Lukin, Head of the Antarctic Expedition told *Russian Reporter* magazine.

...

Looking for Hitler's secret base

On a more far-fetched note, the breakthrough has given new hope to those searching for a secret base built by the Nazis in the ice cover of the same lake.

Grand Admiral Karl Dontiz told Adolf Hitler in 1943 that "Germany's submarine fleet is proud that it created an unassailable fortress for the Fuhrer on the other end of the world." It was even speculated that the remains of Hitler and his wife Eva Braun were transported to the base. Perhaps, the mission will now discover the base near the lake, though this appears somewhat unlikely.

But even less spectacular discoveries will have to wait—the Antarctic summer is coming to an end, and the

> expedition will have to return to warmer climes. Russian scientists are proposing to use specially-designed swimming robots to collect the first samples from the lake water at the end of the year.

Bringing up Hitler and Nazis in Antarctica during WWII, and that they may have had a base near Lake Vostok, seems like a very strange tack to take in an article concerning breaking through to Lake Vostok. What was the purpose of even mentioning such a thing in an article about a frozen lake beneath the Antarctic ice sheet? Is it just a reference to popular beliefs about Antarctica included in order to jazz up a boring story about how only the moon of Europa has a similar climate?

An article just as strange appeared on the Fox News website on February 3, 2012, two days before the announcement of having successfully drilled into Lake Vostok. The title of the article was: "Missing scientists mystery deepens in frozen Antarctica." The article read:

> The world holds its breath, hoping for the best after six days of radio silence from Antarctica—where a team of Russian scientists is racing the clock and the oncoming winter to dig to an alien lake far beneath the ice. The team from Russia's Arctic and Antarctic Research Institute (AARI) have been drilling for weeks in an effort to reach isolated Lake Vostok, a vast, dark body of water hidden 13,000 ft. below the surface of the icy continent. Lake Vostok hasn't been exposed to air in more than 20 million years.
>
> The team's last contact with colleagues in the unfrozen world was six long days ago, and scientists from around the globe are unsure of the fate of the mission—and the scientists themselves—as Antarctica's killing winter draws near.
>
> "When you're outside, it's extremely cold—minus 30, minus 40," microbiologist Dr. David A. Pearce told FoxNews.com. "If you left your eyes open the fluid in them would start to freeze. Your nostrils would start to freeze. The moisture in your mouth would start to freeze," he said.
>
> Pearce heads a team from the British Antarctic Survey

on a competing mission, set to plumb the depths of Lake Ellsworth, one of a string of more than 370 lakes beneath Antarctica that may soon see light for the first time since well before Fred Flintstone's ancestors roamed the planet. But time is running out for the Russian scientists.

"They need to be out by the 6th of February," Pearce said, when winter sets in and temperatures drop another 40 degrees centigrade. Vostok Station boasts the lowest recorded temperature on Earth: -129 degrees Fahrenheit (-89.4 degrees Celsius).

The Russian scientists have been communicating with Pearce and colleagues at a third Antarctic expedition—a study of the subglacial Whillans Ice Stream mainly featuring U.S. scientists. The competing teams have been watching the Russians and sharing notes over the past few days, Pearce told FoxNews.com—yet no one knows what has happened.

"We're all waiting with bated breath," he said.

...The Russian Arctic and Antarctic Research Institute did not respond to FoxNews.com requests for information on the status of the team, information that did not surprise Priscu.

"The Russians have their own way of dealing with things, particularly the media, which I respect," he told FoxNews.com. "There is nobody to call."

Other science institutions that typically conduct investigations in Antarctica did not respond to requests for information, including the National Oceanic and Atmospheric Administration (NOAA), the NSF's United States Antarctic Program, or the U.S. Ice Drilling Program.

"They are running out of time to complete the drilling this year," Pearce said.

Two days later the Russians announced that they had completed drilling into Lake Vostok and then mentioned the Nazi base rumored to be somewhere around Lake Vostok. Had they found anything else in the week of radio silence? No more was heard of "missing scientists" and apparently everyone at Vostok Station was fine. They apparently left the station on February 8 of that year.

That Russian and American scientists, astronauts and military personnel are cooperating in Antarctica—as well as in space with the International Space Station—is an everyday fact. Yet, politicians in both countries like to maintain that the two countries are adversaries, jockeying for prime positions on the global scene. Yet, while we rarely hear news from the Amundsen–Scott South Pole Station (anyone seen a UFO lately?) we are reminded that the Russians are our friends when they launch our own NASA astronauts to the International Space Station in a Russian-made rocket at a Russian-made launch site.

Remember that Gary McKinnon said that he saw documents for Solar Warden that said other countries were part of this international space force, including Russia, Austria, Australia and the UK. One would not naturally think that Russia would be part of Solar Warden but it actually makes perfect sense. The Russians and the Americans have been working secretly together since their public alliance during WWII. They work together in Antarctica and they work together in space. In light of this we can see how bases on the Moon—and Mars—are also likely to be shared bases with Russian personnel. According to Gary McKinnon, we also have Canadians, Australians, British and others at these bases. We can largely assume that all of these personnel are military officers of one kind or another. We know that there are over 100 US Navy officers in space—Gary McKinnon saw a list of these officers—but now we know that officers from other countries are also in these spaceships and inside these bases.

What is the mission then of Solar Warden and the secret space program? Is it to discover new worlds, as in the *Star Trek* television series? Is it to keep us safe from nasty extraterrestrials who want our planet for their own? Do the Germans, as a Third Power and a breakaway civilization, have their own space fleet? Is there still a base for "submarines that can fly" somewhere in the South Shetland Islands off the coast of Antarctica?

We see on an ongoing basis that UFO activity continues in South America at a fairly steady rate. Perhaps the activity of the Germans will continue until the Western democracies and the oil companies come clean as to the suppressed technology that has been in use since the last months of WWII. Henry Stevens thinks that the renegade hierarchy of the Third Power finally died off in the 1970s and their activities were integrated into the American space program during the 1980s. At the end of this decade came the dissolution of the Soviet Union and suddenly Russia wasn't a Communist country anymore. However, they were still our "adversary." It seems that the integrated space program at this time was between the Russian, German and American secret space programs. A base on the Moon may have already been shared with the Russians by 1966.

Leaving the Blue Avians and the Reptoids aside, it seems that the big question with Antarctica—as well as South America—and UFO activity is why it is still being seen there? Would we be seeing clandestine UFO activity like this if the German breakaway civilization has truly been fully integrated into the US Navy's Solar Warden space program? It would appear that it has not. It seems the German breakaway civilization, now commanded by Volga Germans and other German-South Americans who were only born in the 1960s and 1970s is still functioning to this day.

It seems unlikely that Neuschwabenland is still an active area for the Germans, but their base in the South Shetland Islands may well still be active, if hidden well enough. This group would have their major bases in South America and their children would go to school in South America. They might still have secret bases on remote jungle mountaintops in places such as in New Guinea or the Solomon Islands. And, they probably still have contacts back in

A typical coastline in the South Shetland Islands.

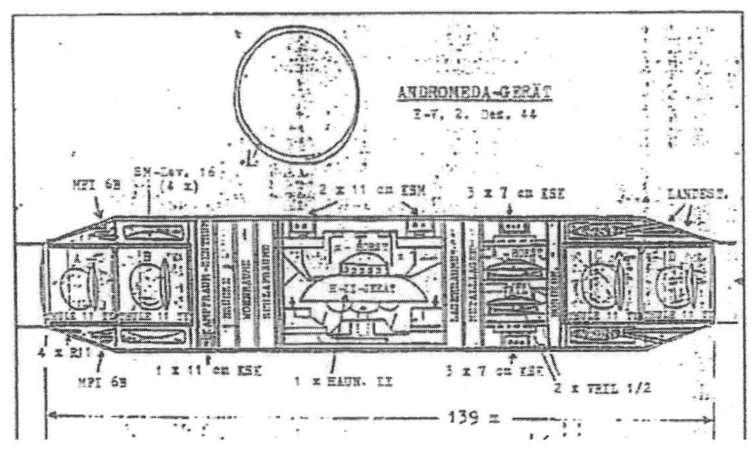

The astonishing plans for the Andromeda Craft—a flying submarine.

Germany, Switzerland and Austria.

One curious incident to mention in this regard was the mass UFO sighting on New Year's Eve in the island nation of Tonga in the year 2000. On that night, Tonga was the first nation in the world to see New Year's Day of the year 2000, widely regarded as the first day of the new millennium, although that landmark would not take place until January 1, 2001. There were many parties all over the main island. Restaurant- and party-goers on the west side of the island all saw a gigantic cigar-shaped craft hovering silently and glowing a few miles off shore, apparently watching the festivities. The craft hovered in the distance over the ocean for more than an hour while hundreds of people gazed at the unusual sight in the sky, and continued to drink and party like it was 1999. Eventually the craft departed and the party-goers went home to a new year. I was in Tonga shortly after this happened and it was the talk of the island.

Is it possible that this craft is one of the Andromeda craft that came from the German documents? If so, maybe this craft came from a secret base in the Solomon Islands or New Guinea, further to the west of Tonga. In fact, the huge spaceship might have even come from Antarctica.

So, in summing things up, I must remark on how much the subject of Ufology has changed in the last 50 years. It started during WWII with the sighting of glowing balls over Germany and then flying saucers. Kenneth Arnold saw a flying wing identical to the German Horton flying wing "skipping like a saucer" over Washington State in 1947. Early reports of flying saucers suggested that they were possibly of German origin, but then the Roswell crash happened and film and print media began to flood the public with a message that there was an alien invasion, and that it was Earth versus the Flying Saucers. These flying saucers held aliens who were here to take over the Earth—or something like that.

This take on reality held throughout the 1960s and 1970s and UFO groups and magazines earnestly studied the Men in Black, flying saucers around the Apollo missions, and such. But as the 1980s came on, cracks in this façade were beginning to show. The Apollo missions to the Moon had come to a complete end. Diagrams and patents of flying saucers and anti-gravity aircraft were published, and the great scientist Nikola Tesla became recognized as the genius that he was after being suppressed since WWII, during which he died.

Then the document dump by Ralf Ettl in 1989, as revealed by Henry Stevens, changed the landscape in Europe, the United States and elsewhere, though it took years for them to filter into books and documentaries. While some diagrams and photos had appeared before, such as those in works by Michael X. Barton,[48] Renato Vesco[69] and others, these new documents were revelatory of what many had suspected and written about for years. Here was genuine proof of the historical designing and manufacturing of flying saucers (discoid craft) and cigar-shaped craft. While this does not explain all UFO incidents in any way, it does provide a framework to look at certain curious UFO incidents and make some sense of them, as I have tried to do in this book.

I do not think that these German documents and other evidence are the explanation for everything in the UFO universe. By no means does this explain all of the many UFO events prior to WWII nor many of the odd occurrences that have been reported year after year around the world in the ensuing 70 years. But, it does help us make some sense out of some of the incidents that have been discussed in this book.

All I can really say is that we live on a strange planet in a strange universe, where the weirdness factor jumps from zero to ten in a microsecond. As Mark Twain said, "It's no wonder that truth is stranger than fiction. Fiction has to make sense." Let's leave it at that.

FOOTNOTES & BIBLIOGRAPHY

1. *The Ratline*, Peter Levenda, 2012, Ibis Press, Lakeworth, FL.
2. *Hess and the Penguins,* Joseph P. Farrell, 2017, Adventures Unlimited Press, Kempton, IL.
3. *The Assassination of James Forrestal*, David Martin, 2019, McCabe Publishing, Hyattsville, MD.
4. *Ice Is Where You Find It*, Capt. Charles Thomas, 1951, Bob-Merrill, New York.
5. *Assassinations: The Plots, Politics, and Powers Behind History-Changing Murders*, Nick Redfern, 2020, Visible Ink Press, Detroit, MI.
6. *Type VII U-boats*, Robert Stern, 1991, Arms and Armor Press, London.
7. *UFOs and Nukes*, Robert Hastings, 2017, (self-published on Amazon).
8. *German Secret Weapons of the Second World War,* Ian V. Hogg, 1999, Stackpole Books, Mechanicsburg, PA.
9. *Donitz and the Wolf Packs*, Bernard Edwards, 1999, Hodder Headline, London.
10. *Maps of the Ancient Sea Kings*, Charles Hapgood, 1966, (reprinted 1997, Adventures Unlimited Press, Kempton, IL).
11. *The U-Boat Wars*, Edwin P. Hoyt, 1984, Stein and Day Publishers, New York.
12. *Nazi International*, Joseph P. Farrell, 2008, Adventures Unlimited Press, Kempton, IL.
13. *The SS Totenkopf Ring,* Craig Gottlieb, 2008, Schiffer Publishing, Atglen, PA.
14. *Hidden Agenda,* Mike Bara, 2016, Adventures Unlimited Press, Kempton, IL.
15. *Vimana*, David Hatcher Childress, 2013, Adventures Unlimited Press, Kempton, IL.
16. *The Hollow Earth*, Dr. Raymond Bernard, 1964, (reprinted 2007, AUP, Kempton, IL).
17. *Subterranean Worlds*, Walter Kafton-Minkel, 1989, Loompanics, Port Townsend, WA.

18. *Agharta: The Subterranean World,* Dr. Raymond Bernard, 1959, Privately published in Brazil.
19. *Flying Saucers from the Earth's Interior*, Dr. Raymond Bernard, 1960, Privately published in Brazil.
20. *Worlds Beyond the Poles*, F. Amadeo Giannini, 1959, Vantage Press, New York.
21. *Expedition Fawcett* (published in the U.S. as *Lost Trails, Lost Cities*), Brian Fawcett, 1953, Hutchinson & Co., London.
22. *Mysteries of Ancient South America*, Harold Wilkins, 1946, Citadel Books, New York.
23. *Journey to the Earth's Interior*, Marshall B. Gardner, 1913 (second edition 1920), New York.
24. *Phantom of the Poles*, William Reed, 1906, Walter S. Rockney Company, New York.
25. *The Subterranean World to the Sky: Flying Saucers*, O.C. Huguenin, 1957, Rio de Janeiro.
26. *They Knew Too Much about Flying Saucers,* Gray Barker, 1956, University Books, NY.
27. *Lost Continents and the Hollow Earth*, Richard Shaver and David Hatcher Childress, 1999, Adventures Unlimited Press, Kempton, IL.
28. *Black Sun: Aryan Cults, Esoteric Nazism, and the Politics of Identity*, Nicholas Goodrick-Clarke, 2002, New York University Press, NY.
29. *The Morning of the Magicians*, Jacques Bergier and Louis Pauwels, 1960, 1963 English edition, Stein & Day, New York.
30. *Reich of the Black Sun,* Joseph P. Farrell, 2009, AUP, Kempton, IL
31. *Casebook on the Men in Black*, Jim Keith, 1997, AUP, Kempton, IL.
32. *Remarkable Luminous Phenomena in Nature*, William Corliss, 2001, Sourcebook Project, Glen Arm, MD.
33. *Hitler: The Survival Myth*, Donald M. McKale, 1983, Stein & Day, New York.
34. *Hitler's Terror Weapons: From VI to Vimana*, Geofrey Brooks, 2002, Pen and Sword Books, Barnsley, UK.
35. *Arktos: The Polar Myth*, Joscelyn Godwin, 1996, AUP, Kempton, IL.
36. *Electric UFOs*, Albert Budden, 1998, Blandford Books, London.
37. *Himmler's Crusade: The Nazi Expedition to Find the Origins of the Aryan Race*, Christopher Hale, 2003, John Wiley & Sons, Hoboken, NJ.
38. *Treasures of the Lost Races*, Rene Noorbergen, 1982, Bobbs-Merrill Company, New York.
39. *Roswell and the Reich,* Joseph P. Farrell, 2010, AUP, Kempton, IL
40. *The CIA UFO Papers,* Dan Wright, 2019, MUFON-Red Wheel-Weiser, Newburyport, MA.

41. *Underground Bases and Tunnels*, Richard Sauder, 1997, AUP, Kempton, IL.
42. *Underwater and Underground Bases*, Richard Sauder, 2001, AUP, Kempton, IL.
43. *U-Boat 977: The U-Boat That Escaped to Argentina*, Heinz Schäffer, 1952, Naval Institute Press, Norfolk, VA (First published in Germany in 1952 as *U-977 – 66 Tage unter Wasser*).
44. *Adolf Hitler and the Secrets of the Holy Lance*, Buechner and Bernhart, 1988, Thunderbird Press, Metairie, LA.
45. *Hitler's Ashes*, Buechner and Bernhart, 1989, Thunderbird Press, Metairie, LA.
46. *Dark Star*, Henry Stevens, 2011, Adventures Unlimited Press, Kempton, IL.
48. *The German Saucer Story*, Michael Barton, 1960, Future Press, Los Angeles.
49. *We Want You: Is Hitler Still Alive?,* Michael Barton, 1960, Future Press, Los Angeles.
50. *Emerald Cup—Ark of Gold*, Col. Howard Buechner, 1991, Thunderbird Press, Metairie, LA.
51. *Il Deutsche Flugscheiben und U-Boote Ueberwachen Die Weltmeere,* O. Bergmann, 1989, Hugin Publishing, Germany.
52. *Anti-Gravity & the World Grid*, David Hatcher Childress, 1987, AUP, Kempton, IL.
53. *Vril, The Power of the Coming Race*, Sir Edward Bulwer-Lytton, 1871, Blackwood and Sons, London.
54. *Laserbeams from Star Cities*, Robyn Collins, 1971, Sphere Books, London.
55. *Secret Cities of Old South America*, Harold Wilkins, 1952, London, (reprinted 1998, Adventures Unlimited Press, Kempton, IL).
56. *War in Ancient India*, V. R. Dikshitar, 1944, Oxford University Press (1987 edition published by Motilal Banarsidass, Delhi).
57. *Invisible Residents*, Ivan T. Sanderson, 1970, AUP, Kempton, IL.
58. *The Philadelphia Experiment*, William Moore and Charles Berlitz, 1979, Grosset & Dunlap, New York.
59. *UFO Photographs Around the World*, Edited by Wendell Stevens, 1986, UFO Photo Archives, Tucson, AZ.
60. *Vimana: Flying Machines of the Ancients*, David Hatcher Childress, 2013, AUP, Kempton, IL.
61. *Investigating the Unexplained*, Ivan T. Sanderson, 1972, Prentice Hall, Englewood Cliffs, NJ.
62. *Missing 411: Off the Grid*, David Paulides, 2017, Canamissing

Publishing, Pueblo, CO.
63. *Pole to Pole*, Michael Palin, 1992, Weidenfeld & Nicolson, London.
64. *Dark Fleet*, Len Kasten, 2020, Bear & Company, Rochester, VT.
65. *Antarctica's Hidden History*, Dr. Michael Salla, 2018, Exopolitics Consultants, Pahoa, Hawaii.
66. *US Air Force Secret Space Program*, Dr. Michael Salla, 2019, Exopolitics Consultants, Pahoa, Hawaii.
67. *Selected by Extraterrestrials*, William Tompkins, 2015, Createspace.
68. *Genesis for the Space Race*, 2014, John B. Leith, Timestream Books, Ooltewah, TN.
69. *Man-Made UFOs,* Renato Vesco and David Childress, 1994, AUP, Kempton, IL.
70. *The Enigma of Cranial Deformation,* David Childress and Brien Foerster, 2012, AUP, Kempton, IL.

Get these fascinating books from your nearest bookstore or directly from:
Adventures Unlimited Press
www.adventuresunlimitedpress.com

COVERT WARS & BREAKAWAY CIVILIZATIONS
By Joseph P. Farrell

Farrell delves into the creation of breakaway civilizations by the Nazis in South America and other parts of the world. He discusses the advanced technology that they took with them at the end of the war and the psychological war that they waged for decades on America and NATO. He investigates the secret space programs currently sponsored by the breakaway civilizations and the current militaries in control of planet Earth. Plenty of astounding accounts, documents and speculation on the incredible alternative history of hidden conflicts and secret space programs that began when World War II officially "ended."

292 Pages. 6x9 Paperback. Illustrated. $19.95. Code: BCCW

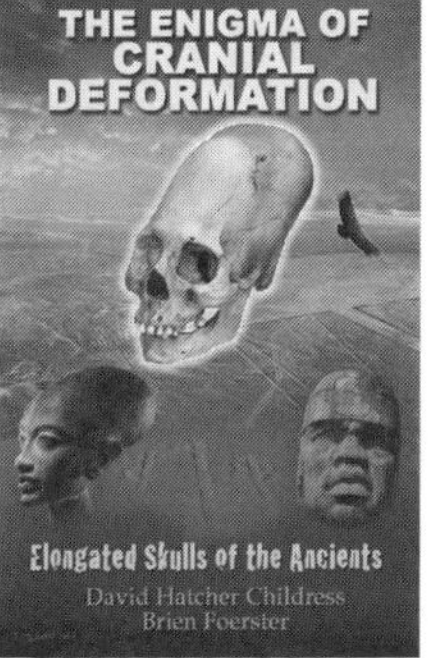

THE ENIGMA OF CRANIAL DEFORMATION
Elongated Skulls of the Ancients
By David Hatcher Childress and Brien Foerster

In a book filled with over a hundred astonishing photos and a color photo section, Childress and Foerster take us to Peru, Bolivia, Egypt, Malta, China, Mexico and other places in search of strange elongated skulls and other cranial deformation. The puzzle of why diverse ancient people—even on remote Pacific Islands—would use head-binding to create elongated heads is mystifying. Where did they even get this idea? Did some people naturally look this way—with long narrow heads? Were they some alien race? Were they an elite race that roamed the entire planet? Why do anthropologists rarely talk about cranial deformation and know so little about it? Color Section.

250 Pages. 6x9 Paperback. Illustrated. $19.95. Code: ECD

ARK OF GOD
The Incredible Power of the Ark of the Covenant
By David Hatcher Childress

Childress takes us on an incredible journey in search of the truth about (and science behind) the fantastic biblical artifact known as the Ark of the Covenant. This object made by Moses at Mount Sinai—part wooden-metal box and part golden statue—had the power to create "lightning" to kill people, and also to fly and lead people through the wilderness. The Ark of the Covenant suddenly disappears from the Bible record and what happened to it is not mentioned. Was it hidden in the underground passages of King Solomon's temple and later discovered by the Knights Templar? Was it taken through Egypt to Ethiopia as many Coptic Christians believe? Childress looks into hidden history, astonishing ancient technology, and a 3,000-year-old mystery that continues to fascinate millions of people today. Color section.

420 Pages. 6x9 Paperback. Illustrated. $22.00 Code: AOG

YETIS, SASQUATCH & HAIRY GIANTS
By David Hatcher Childress

Childress takes the reader on a fantastic journey across the Himalayas to Europe and North America in his quest for Yeti, Sasquatch and Hairy Giants. Childress begins with a discussion of giants and then tells of his own decades-long quest for the Yeti in Nepal, Sikkim, Bhutan and other areas of the Himalayas, and then proceeds to his research into Bigfoot, Sasquatch and Skunk Apes in North America. Chapters include: The Giants of Yore; Giants Among Us; Wildmen and Hairy Giants; The Call of the Yeti; Kanchenjunga Demons; The Yeti of Tibet, Mongolia & Russia; Bigfoot & the Grassman; Sasquatch Rules the Forest; Modern Sasquatch Accounts; more. Includes a 16-page color photo insert of astonishing photos!

360 pages. 5x9 Paperback. Illustrated. Bibliography. Index. $18.95. Code: YSHG

SECRETS OF THE HOLY LANCE
The Spear of Destiny in History & Legend
by Jerry E. Smith

Secrets of the Holy Lance traces the Spear from its possession by Constantine, Rome's first Christian Caesar, to Charlemagne's claim that with it he ruled the Holy Roman Empire by Divine Right, and on through two thousand years of kings and emperors, until it came within Hitler's grasp—and beyond! Did it rest for a while in Antarctic ice? Is it now hidden in Europe, awaiting the next person to claim its awesome power? Neither debunking nor worshiping, *Secrets of the Holy Lance* seeks to pierce the veil of myth and mystery around the Spear.

312 PAGES. 6x9 PAPERBACK. ILLUSTRATED. $16.95. CODE: SOHL

THE CRYSTAL SKULLS
Astonishing Portals to Man's Past
by David Hatcher Childress and Stephen S. Mehler

Childress introduces the technology and lore of crystals, and then plunges into the turbulent times of the Mexican Revolution form the backdrop for the rollicking adventures of Ambrose Bierce, the renowned journalist who went missing in the jungles in 1913, and F.A. Mitchell-Hedges, the notorious adventurer who emerged from the jungles with the most famous of the crystal skulls. Mehler shares his extensive knowledge of and experience with crystal skulls. Having been involved in the field since the 1980s, he has personally examined many of the most influential skulls, and has worked with the leaders in crystal skull research. Color section.

294 pages. 6x9 Paperback. Illustrated. $18.95. Code: CRSK

THE LAND OF OSIRIS
An Introduction to Khemitology
by Stephen S. Mehler

Was there an advanced prehistoric civilization in ancient Egypt? Were they the people who built the great pyramids and carved the Great Sphinx? Did the pyramids serve as energy devices and not as tombs for kings? Chapters include: Egyptology and Its Paradigms; Khemitology—New Paradigms; Asgat Nefer—The Harmony of Water; Khemit and the Myth of Atlantis; The Extraterrestrial Question; more. Color section.

272 PAGES. 6x9 PAPERBACK. ILLUSTRATED . $18.95. CODE: LOOS

VIMANA:
Flying Machines of the Ancients
by David Hatcher Childress

According to early Sanskrit texts the ancients had several types of airships called vimanas. Like aircraft of today, vimanas were used to fly through the air from city to city; to conduct aerial surveys of uncharted lands; and as delivery vehicles for awesome weapons. David Hatcher Childress, popular *Lost Cities* author, takes us on an astounding investigation into tales of ancient flying machines. In his new book, packed with photos and diagrams, he consults ancient texts and modern stories and presents astonishing evidence that aircraft, similar to the ones we use today, were used thousands of years ago in India, Sumeria, China and other countries. Includes a 24-page color section.

408 Pages. 6x9 Paperback. Illustrated. $22.95. Code: VMA

THE LOST WORLD OF CHAM
The Trans-Pacific Voyages of the Champa
By David Hatcher Childress

The mysterious Cham, or Champa, peoples of Southeast Asia formed a megalith-building, seagoing empire that extended into Indonesia, Tonga, and beyond—a transoceanic power that reached Mexico and South America. The Champa maintained many ports in what is today Vietnam, Cambodia, and Indonesia and their ships plied the Indian Ocean and the Pacific, bringing Chinese, African and Indian traders to far off lands, including Olmec ports on the Pacific Coast of Central America. Topics include: Cham and Khem: Egyptian Influence on Cham; The Search for Metals; The Basalt City of Nan Madol; Elephants and Buddhists in North America; The Cham and Lake Titicaca; Easter Island and the Cham; the Magical Technology of the Cham; tons more. 24-page color section.

328 Pages. 6x9 Paperback. Illustrated. $22.00 Code: LPWC

ADVENTURES OF A HASHISH SMUGGLER
by Henri de Monfreid

Nobleman, writer, adventurer and inspiration for the swashbuckling gun runner in the *Adventures of Tintin*, Henri de Monfreid lived by his own account "a rich, restless, magnificent life" as one of the great travelers of his or any age. The son of a French artist who knew Paul Gaugin as a child, de Monfreid sought his fortune by becoming a collector and merchant of the fabled Persian Gulf pearls. He was then drawn into the shadowy world of arms trading, slavery, smuggling and drugs. Infamous as well as famous, his name is inextricably linked to the Red Sea and the raffish ports between Suez and Aden in the early years of the twentieth century. De Monfreid (1879 to 1974) had a long life of many adventures around the Horn of Africa where he dodged pirates as well as the authorities.

284 Pages. 6x9 Paperback. $16.95. Illustrated. Code AHS

NORTH CAUCASUS DOLMENS
In Search of Wonders
By Boris Loza, Ph.D.

Join Boris Loza as he travels to his ancestral homeland to uncover and explore dolmens firsthand. Throughout this journey, you will discover the often hidden, and surprisingly forbidden, perspective about the mysterious dolmens: their ancient powers of fertility, healing and spiritual connection. Chapters include: Ancient Mystic Megaliths; Who Built the Dolmens?; Why the Dolmens were Built; Asian Connection; Indian Connection; Greek Connection; Olmec and Maya Connection; Sun Worshippers; Dolmens and Archeoastronomy; Location of Dolmen Quarries; Hidden Power of Dolmens; and much more! Tons of Illustrations! A fascinating book of little-seen megaliths. Color section.

252 Pages. 5x9 Paperback. Illustrated. $24.00. Code NCD

THE ENCYCLOPEDIA OF MOON MYSTERIES
Secrets, Anomalies, Extraterrestrials and More
By Constance Victoria Briggs

Our moon is an enigma. The ancients viewed it as a light to guide them in the darkness, and a god to be worshipped. Some even believe that there are cities beneath the surface of the Moon. Did you know that: Aristotle and Plato wrote about a time when there was no Moon? Several of the NASA astronauts reported seeing UFOs while traveling to the Moon?; the Moon might be hollow?; Apollo 10 astronauts heard strange "space music" when traveling on the far side of the Moon?; strange and unexplained lights have been seen on the Moon for centuries?; there are said to be ruins of structures on the Moon?; there is an ancient tale that suggests that the first human was created on the Moon?; Tons more. Tons of illustrations with A to Z sections for easy reference and reading.

152 Pages. 7x10 Paperback. Illustrated. $19.95. Code: EOMM

TECHNOLOGY OF THE GODS
The Incredible Sciences of the Ancients
by David Hatcher Childress

Childress looks at the technology that was allegedly used in Atlantis and the theory that the Great Pyramid of Egypt was originally a gigantic power station. He examines tales of ancient flight and the technology that it involved; how the ancients used electricity; megalithic building techniques; the use of crystal lenses and the fire from the gods; evidence of various high tech weapons in the past, including atomic weapons; ancient metallurgy and heavy machinery; the role of modern inventors such as Nikola Tesla in bringing ancient technology back into modern use; impossible artifacts; and more.

356 pages. 6x9 Paperback. Illustrated. $16.95. code: TGOD

THE ANTI-GRAVITY HANDBOOK
edited by David Hatcher Childress

The new expanded compilation of material on Anti-Gravity, Free Energy, Flying Saucer Propulsion, UFOs, Suppressed Technology, NASA Cover-ups and more. Highly illustrated with patents, technical illustrations and photos. This revised and expanded edition has more material, including photos of Area 51, Nevada, the government's secret testing facility. This classic on weird science is back in a new format!

230 PAGES. 7X10 PAPERBACK. ILLUSTRATED. $16.95. CODE: AGH

ANTI–GRAVITY & THE WORLD GRID

Is the earth surrounded by an intricate electromagnetic grid network offering free energy? This compilation of material on ley lines and world power points contains chapters on the geography, mathematics, and light harmonics of the earth grid. Learn the purpose of ley lines and ancient megalithic structures located on the grid. Discover how the grid made the Philadelphia Experiment possible. Explore the Coral Castle and many other mysteries, including acoustic levitation, Tesla Shields and scalar wave weaponry. Browse through the section on anti-gravity patents, and research resources.

274 PAGES. 7X10 PAPERBACK. ILLUSTRATED. $14.95. CODE: AGW

ANTI–GRAVITY & THE UNIFIED FIELD
edited by David Hatcher Childress

Is Einstein's Unified Field Theory the answer to all of our energy problems? Explored in this compilation of material is how gravity, electricity and magnetism manifest from a unified field around us. Why artificial gravity is possible; secrets of UFO propulsion; free energy; Nikola Tesla and anti-gravity airships of the 20s and 30s; flying saucers as superconducting whirls of plasma; anti-mass generators; vortex propulsion; suppressed technology; government cover-ups; gravitational pulse drive; spacecraft & more.

240 PAGES. 7X10 PAPERBACK. ILLUSTRATED. $14.95. CODE: AGU

THE TIME TRAVEL HANDBOOK
A Manual of Practical Teleportation & Time Travel
edited by David Hatcher Childress

The Time Travel Handbook takes the reader beyond the government experiments and deep into the uncharted territory of early time travellers such as Nikola Tesla and Guglielmo Marconi and their alleged time travel experiments, as well as the Wilson Brothers of EMI and their connection to the Philadelphia Experiment—the U.S. Navy's forays into invisibility, time travel, and teleportation. Childress looks into the claims of time travelling individuals, and investigates the unusual claim that the pyramids on Mars were built in the future and sent back in time. A highly visual, large format book, with patents, photos and schematics. Be the first on your block to build your own time travel device!

316 PAGES. 7X10 PAPERBACK. ILLUSTRATED. $16.95. CODE: TTH

OBELISKS: TOWERS OF POWER
The Mysterious Purpose of Obelisks
By David Hatcher Childress

Some obelisks weigh over 500 tons and are massive blocks of polished granite that would be extremely difficult to quarry and erect even with modern equipment. Why did ancient civilizations in Egypt, Ethiopia and elsewhere undertake the massive enterprise it would have been to erect a single obelisk, much less dozens of them? Were they energy towers that could receive or transmit energy? With discussions on Tesla's wireless power, and the use of obelisks as gigantic acupuncture needles for earth, Chapters include: Megaliths Around the World and their Purpose; The Crystal Towers of Egypt; The Obelisks of Ethiopia; Obelisks in Europe and Asia; Mysterious Obelisks in the Americas; The Terrible Crystal Towers of Atlantis; Tesla's Wireless Power Distribution System; Obelisks on the Moon; more. 8-page color section.

336 Pages. 6x9 Paperback. Illustrated. $22.00 Code: OBK

NIKOLA TESLA'S ELECTRICITY UNPLUGGED
Wireless Transmission of Power as the Master of Lightning Intended
Edited by Tom Valone, Ph.D.

The immense genius of Tesla resulted from his ability to see an invention in 3-D, from every angle, within his mind before it was easily built. Tesla's inventions were complete down to dimensions and part sizes in his visionary process. Tesla would envision his electromagnetic devices as he stared into the sky, or into a corner of his laboratory. His inventions on rotating magnetic fields, creating AC current as we know it today, have changed the world—yet most people have never heard of this great inventor. Includes: Tesla's fantastic vision of the future, his wireless transmission of power, Tesla's Magnifying Transmitter, the testing and building of his towers for wireless power, tons more. The genius of Nikola Tesla is being realized by millions all over the world!

464 pages. 6x9 Paperback. Illustrated. Index. $21.95 Code: NTEU

THE TESLA PAPERS
Nikola Tesla on Free Energy & Wireless Transmission of Power
by Nikola Tesla, edited by David Hatcher Childress

David Hatcher Childress takes us into the incredible world of Nikola Tesla and his amazing inventions. Tesla's fantastic vision of the future, including wireless power, anti-gravity, free energy and highly advanced solar power. Also included are some of the papers, patents and material collected on Tesla at the Colorado Springs Tesla Symposiums, including papers on: •The Secret History of Wireless Transmission •Tesla and the Magnifying Transmitter •Design and Construction of a Half-Wave Tesla Coil •Electrostatics: A Key to Free Energy •Progress in Zero-Point Energy Research •Electromagnetic Energy from Antennas to Atoms

325 PAGES. 8X10 PAPERBACK. ILLUSTRATED. $16.95. CODE: TTP

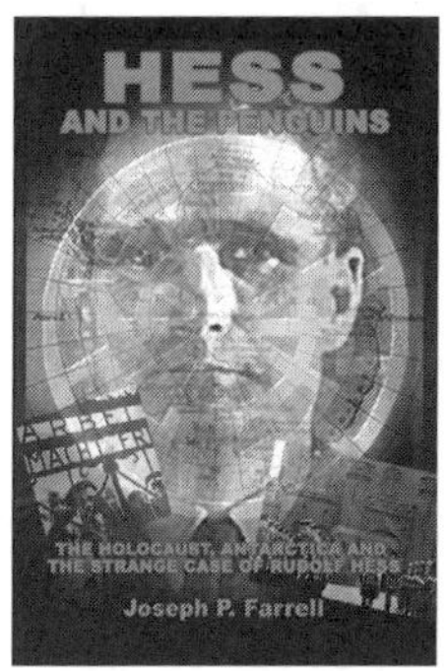

HESS AND THE PENGUINS
The Holocaust, Antarctica and the Strange Case of Rudolf Hess
By Joseph P. Farrell

Farrell looks at Hess' mission to make peace with Britain and get rid of Hitler—even a plot to fly Hitler to Britain for capture! How much did Göring and Hitler know of Rudolf Hess' subversive plot, and what happened to Hess? Why was a doppleganger put in Spandau Prison and then "suicided"? Did the British use an early form of mind control on Hess' double? John Foster Dulles of the OSS and CIA suspected as much. Farrell also uncovers the strange death of Admiral Richard Byrd's son in 1988, about the same time of the death of Hess.

288 Pages. 6x9 Paperback. Illustrated. $19.95. Code: HAPG

ORDER FORM

10% Discount When You Order 3 or More Items!

One Adventure Place
P.O. Box 74
Kempton, Illinois 60946
United States of America
Tel.: 815-253-6390 • Fax: 815-253-6300
Email: auphq@frontiernet.net
http://www.adventuresunlimitedpress.com

ORDERING INSTRUCTIONS

✓ Remit by USD$ Check, Money Order or Credit Card
✓ Visa, Master Card, Discover & AmEx Accepted
✓ Paypal Payments Can Be Made To: info@wexclub.com
✓ Prices May Change Without Notice
✓ 10% Discount for 3 or More Items

SHIPPING CHARGES

United States

✓ Postal Book Rate { $4.50 First Item; 50¢ Each Additional Item
✓ POSTAL BOOK RATE Cannot Be Tracked! Not responsible for non-delivery.
✓ Priority Mail { $7.00 First Item; $2.00 Each Additional Item
✓ UPS { $9.00 First Item (Minimum 5 Books); $1.50 Each Additional Item
NOTE: UPS Delivery Available to Mainland USA Only

Canada

✓ Postal Air Mail { $19.00 First Item; $3.00 Each Additional Item
✓ Personal Checks or Bank Drafts MUST BE US$ and Drawn on a US Bank
✓ Canadian Postal Money Orders OK
✓ Payment MUST BE US$

All Other Countries

✓ Sorry, No Surface Delivery!
✓ Postal Air Mail { $19.00 First Item; $7.00 Each Additional Item
✓ Checks and Money Orders MUST BE US$ and Drawn on a US Bank or branch.
✓ Paypal Payments Can Be Made in US$ To: info@wexclub.com

SPECIAL NOTES

✓ RETAILERS: Standard Discounts Available
✓ BACKORDERS: We Backorder all Out-of-Stock Items Unless Otherwise Requested
✓ PRO FORMA INVOICES: Available on Request
✓ DVD Return Policy: Replace defective DVDs only

ORDER ONLINE AT: www.adventuresunlimitedpress.com

10% Discount When You Order 3 or More Items!

Please check: ✓

☐ This is my first order ☐ I have ordered before

Name
Address
City
State/Province | Postal Code
Country
Phone: Day | Evening
Fax | Email

Item Code	Item Description	Qty	Total

Please check: ✓

☐ Postal-Surface
☐ Postal-Air Mail (Priority in USA)
☐ UPS (Mainland USA only)

Subtotal ▸
Less Discount-10% for 3 or more items ▸
Balance ▸
Illinois Residents 6.25% Sales Tax ▸
Previous Credit ▸
Shipping ▸
Total (check/MO in USD$ only) ▸

☐ Visa/MasterCard/Discover/American Express

Card Number:
Expiration Date: Security Code:

✓ SEND A CATALOG TO A FRIEND: